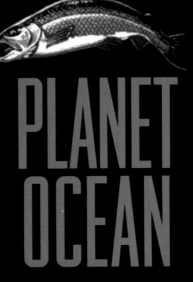

PLANET OCEAN

A Story of Life, the Sea, and Dancing to the Fossil Record

WRITTEN BY BRAD MATSEN
ILLUSTRATED BY RAY TROLL

TEN SPEED PRESS
BERKELEY

For Holly
B.C.M.

For Michelle
R.T.

TEN SPEED PRESS
P. O. Box 7123
Berkeley, CA 94707

Jacket and text designed by Kate Thompson
Edited by Marlene Blessing

Library of Congress
Cataloging-in-Publication Data
Matsen, Bradford.
    Planet Ocean : a story of life, the sea,
  and dancing to the fossil record / by Brad
  Matsen ; illustrations by Ray Troll.
       p.   cm.
    Includes index.
    ISBN 0-89815-778-1
    1. Evolutionary paleobiology.
    2. Fossils—North America I. Title
  QE721.2.E85M38   1994
  560.973—dc20   94-5174
                         CIP

FIRST PRINTING 1995
Printed in Singapore
1 2 3 4 5 — 99 98 97 96 95

> " We shall not cease from exploration
> And the end of all our exploring
> Will be to arrive where we started
> And know the place for the first time. "
>
> —T. S. ELIOT, "FOUR QUARTETS"

PRECAMBRIAN ERA 4,600 - 2,500 MILLION YEARS AGO

CONTENTS

CAMBRIAN EXPLOSION

1. EMERALDELLA
2. ANOMALOCARIS
3. NARAOIA COMPACTA
4. DINOMISCHUS ISOLATUS
5. HALLUCIGENIA SPARSA
6. BRACHIOPOD
7. SAROTROCERCUS OBLITA
8. ECHINODERM
9. ARCHAEOCYATHID
10. LEANCHOILIA SUPERLATA
11. HALKIERIIDS
12. VETULICOLA CUNEATUS
13. TRILOBITE
14. MARRELLA SPLENDENS
15. AYSHEAIA PEDUNCULATA
16. WIWAXIA
17. OTTOIA PROLIFICA
18. NECTOCARIS
19. OPABINIA REGALIS
20. HYOLITHES
21. AMISKWIA
22. BRACHNIA
23. WAPTIA
24. PIKAIA GRACILENS
25. MICRODICTYON SINICUM

Trilobite Safari

SAFARI

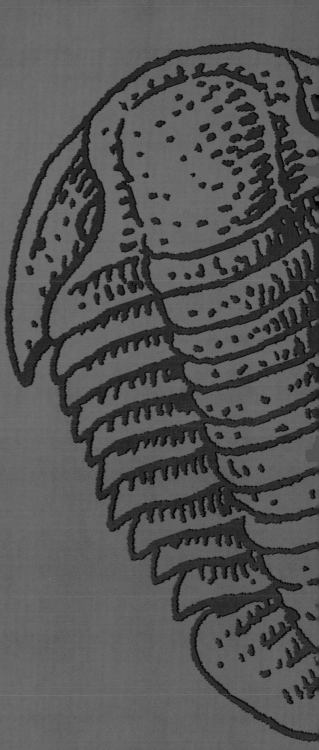

LIFE AS WE KNOW IT

OVERTURE ON FISH CREEK

IN THE BEGINNING, Ray and I were shivering in the bark-brown water of an Alaskan creek in February, coming to terms with the death of a fish. Chip, one of our pals fishing across a gravel bar, hooked a fine steelhead, and we stopped casting, watched the action, and savored the sensual feast around us. The cold air carried the loamy aroma of the Tongass Forest, the stream sang complex harmonies over cobbled bottom on its way to the sea, and the steelhead ploughed a desperate hump in the dark surface of the water. Just as the fish struck, a kingfisher scrackled overhead and eagles clattered around on the banks like irritable gardeners raking up debris. Blue-black ravens the size of spaniels chuckled at us from trees. The fish on Chip's line hit a typical Pacific coast bait rig, a laser-sharp hook buried in a glob of salmon eggs. When not wrapped around sudden death, this bit of trickery is the meal of choice for *Onchorhynchus mykiss*, a.k.a. Hooknose my kiss, the steelhead.

Some of the rich orange globes of life-yet-to-be always wash from the streambed gravel where they were deposited a few months earlier by salmon that had died, poignantly, mere hours after making whoopee. Their rotting bodies surrender carbon, nitrogen, and trace elements to nourish the soil, the ferns, and the trees through the guts of bears, the droppings of birds, and the complex root webs of the forest. In return, ages after glaciers and erosion have crumbled rock to gravel, gravel to sand, a new generation of salmon receive the stream's protection and nourishment—unless another fish, a bear, or a sporting predator intervenes.

That moment on the stream stretched itself over time like a taut drumhead to sound a deep, single note:

Nature is a workshop, not a temple.

Like a lot of people, we were wondering just what is going on here on Planet Ocean.

OPPOSITE: The Kingfisher, the Raven, and the Half-Moon Run (detail)

Ray and I went fishing that February and lucked into a way to begin telling this story of life, the sea, and dancing to the fossil record. Like a lot of people, we were wondering just what is going on here on Planet Ocean, why it is that carbon atoms born in the fires of stars have careened through eons of time to produce Sibelius, rap music, joy, sorrow, child abuse, baseball, art, poetry, television, passion, automobiles, Yo-Yo Ma, money, Republicans, birds, and steelhead.

The fish at hand entered the ranks of the living in the glacial till of that northern stream, spun himself into maturity in the Pacific gyre, and returned, then, to whisper messages from ancient ancestors. His natural predators had been the seals, sea lions, killer whales, and seabirds in the upper tiers of the ocean food web. His prey, since the day he absorbed the sac of his own egg, were the eggs of other fish, grubs, insects, and, finally, herring, squid, and other swimming creatures smaller and slower than he. The steelhead fought furiously to eat and to avoid being eaten. He was aware of nothing but an immediate meal or threat and had no sense of a past or future self. He could not know, for instance, of his noble lineage: that more fish have existed than any other vertebrates since emerging from the primordial soup more than half a billion years ago; that somehow his distant relatives survived many mass dyings in the ocean, at least one of which extinguished 96 percent of all life; or that other fish were the first backboned creatures to colonize the land 350 million years ago. In his rush up the stream, invigorated by the urgings of romance and highly oxygenated water over his gills, he remained, as always, in the good old here and now.

Ray broke my reverie with his shark-smile, baring his teeth and flaring his gill plates. He knows me pretty well, and he could tell I was captivated by the primitive thrill of a fish on the line. My reaction was an undeniable contradiction because for the previous year, after spending most of the half-century of my life in a religious relationship with sportfishing, I stopped believing a being should kill or hurt another being for the fun of it. Something in the eye of a blue marlin I saw die on the deck of a Mexican fishing boat had snatched away the enchantment of a billfish strike on a hot, Pacific afternoon or of a whistling flyline over a trout stream. The beautiful animal turning purple and black at my feet was bound for the Mazatlán dump or the wall of a dentist's office in Houston. In that instant, my vision of the web of life changed. Every living thing must kill to eat, and many animals even torture their prey. Human beings killing for sport, though, now leaves me unsettled. We have more in common than not with a fish, and there's no good reason to consider either of us the superior creature except in the matter of predation. Ray and I can catch and eat a steelhead, but a steelhead can't catch and eat me and Ray.

I'm an extremist on the subject of fishing for sport, but I still fish for a campfire meal or my smoker, and Ray ribs me about that sometimes. We share, though, a deep sense of kinship with that steelhead on Fish Creek and decided nothing was more worth doing than tracking down his story—and our own.

We found out, in the course of our search, that 650 million years ago, give or take a few million years, the sea was mother and father to us all. In the age we call the Precambrian, complex life gained a powerful hold on Planet Ocean after billions of trips around the sun as a cocktail of single-celled algae and bacteria. With time and wild luck, that creative overture ended up as the steelhead, you,

LAND AND SEA DURING THE CAMBRIAN

500 to 570 million years BP

We call the single continent Gondwanaland. During the Cambrian, all life lived in the sea, with the possible exception of some landlubber bacteria. By the end of the period, trilobites were the top critters in the marine food web, and complex life had exploded into enormous diversity. Life-as-we-someday-would-know-it had arrived on Planet Ocean.

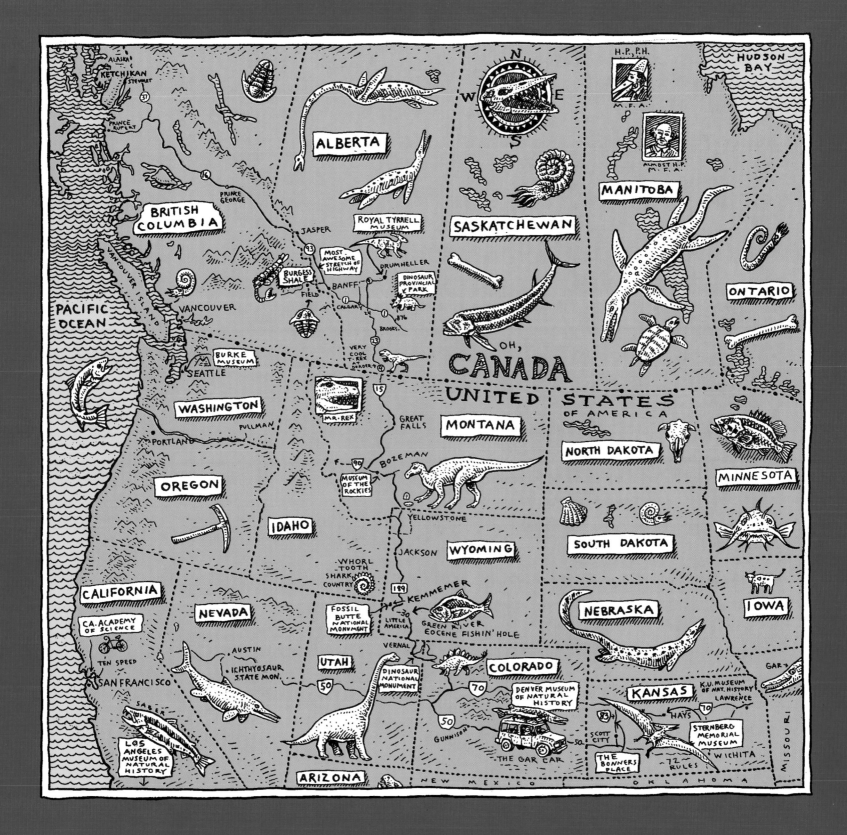

animal body types on Planet Ocean today. Fish and the rest of us backboned creatures have more recent common ancestors, such as the giant, armored swimmers of the early Devonian and the adventurous, or desperate, tribe of lungfish that left the sea to kick off vertebrate life on land. We are of the superkingdom Eukaryota, kingdom Animalia, subkingdom Deuterostomia, phylum Chordata, and subphylum Vertebrata. All that Latin covers 3.5 billion years of evolution, surely enough time to suspect that we are family to a steelhead.

WHY, AFTER FOUR BILLION YEARS . . .

FOSSILS TELL THE STORY of life's journey from its watery beginnings, but because the rocks that carry them are recycled through the tectonic engine, entire chapters are missing, consumed by fire and pressure. Any six-year-old who has played with jigsaw puzzles can look at a globe and understand immediately that Africa just has to fit into South America, Europe into North America, and like that. In fact the continents have been joined and ripped apart several times, and in the late 1960s the dramatic new theory of plate tectonics explained that process and changed forever the way we perceive the ground on which we walk.

Rocks, like living things, arrive and depart, so we have only meager hints about the most distant epochs when complex, multicellular creatures bloomed, eventually to become steelhead and the rest of us. We have even less to go on from the 2 billion or so years when single-celled organisms hovered in the soup of this wannabe paradise, still less from the ancient, organically inert ocean.

With all that uncertainty, it's no wonder folks

me, Ray, the kingfishers, eagles, ravens, and every other living thing.

We also found out that the earliest ancestor we share with that steelhead and every other animal with a backbone was an inch-long long shot named *Pikaia*. The fossil *Pikaia* was lifted into the sky from its moment of death in the ancient sea, some of whose mud turned into the shale of the Canadian Rockies. With a magnifying glass, you can see in it a hint of a spinal cord, the beginnings of phylum Chordata, our tribe, one of the twenty-five or so

FISH · IN · THE · TREES

FISH LADDER TO THE STARS

Fish Ladder
to the Stars

have said life comes from a clamshell, a lush garden, or whatever creation myth they subscribe to. However the deal went down, the real creation event did leave a few clues. We call the age when life exploded the Cambrian, after Cambria—the Romans' name for Wales—because that's where the first rocks of that age were identified. Most remarkably, the Cambrian explosion is kind of the Big Bang of biology, and its main mystery is contained in its brevity. Why, after our home planet formed in a condensing cloud of cosmic dust and rolled through 4.5 billion trips around the sun, did every basic form of life-as-we-know-it suddenly appear within a few million years, along with at least as many other life-forms that have not survived to our time?

It depends on whom you ask.

Sir Fred Hoyle, a British astronomer who incidentally coined the term "Big Bang," and his mathematician sidekick, Chandra Wickramasinghe, say organic microbes are regularly deposited on Planet Ocean by comets and other extraterrestrial visitors. The Cambrian explosion was just one of those instances, they say, and more recent assaults like viruses and influenza are more typical. A lot of religionists will tell you the idea of seeds of life is quite consistent with their version of God-managed creation. All they have to do is tinker with their myths to add more time to the metaphors. Others, fundamental creationists, still say life appeared in a very short time—six days—with all the puzzling lines of evolution as a kind of trick to test their faith. Until the middle of the nineteenth century, the age of earth was a religious, social, and political number, instead of a physical truth. Reformation Christians nailed down not only the year but the date and hour of creation: 9:00 A.M., October 23, 4004 BC. It was

only a few decades ago that we fixed the age of Planet Ocean with radiocarbon dating at about 4.5 billion years.

Life-as-we-would-know-it clearly exploded in the Cambrian, probably because the water and atmosphere of Planet Ocean finally became hospitable to multicellular life. Though the arrival of a complex, diverse web of species appears in the fossil record to have been very abrupt, a lot of less dramatic evolution happened before the Cambrian. In fact, the miracle was really a three-step program. The oldest fossils are microscopic specks of single-cell algae in rare rocks at least 3 billion years old. They tell us that light and photosynthesis were present and record the breakthrough from inorganic to organic molecules. This was the first step in the miracle, almost certainly a piece of chemical luck, which can be duplicated in the simplest of laboratories today. In the fifties, while a lot of us were watching Howdy Doody, eating fish sticks and Velveeta cheese, and voting for Eisenhower, a couple of chemists in Chicago mixed methane, ammonia, hydrogen, and water vapor and zapped them with electricity for a week. A dark scum, vaguely like the earliest algae, formed in their flask, which proved to contain four amino acids, urea, and a few fatty acids. It's a long reach from a dark scum in a flask to Ray Troll or a steelhead, but time, say 3 or 4 billion years, can do the job.

The vital chore the dark scum performed was production of photosynthetic cells that can make oxygen, just like your houseplants. The fossil record, meager though it is from 2 billion years ago, shows clearly that oxygen was present in great quantity because rust appears in the traces of ferrous iron in the rocks. Once single-celled algae got to work making oxygen, you were history if you happened to

The leap from inorganic ditch water to oxygen-making organic scum was the first third of the miracle of life.

colonization of the light-rich shallower waters, and wiped out competing primitive cells, the miracle continued. If you're an organism whose cells lack defined nuclei, you're a prokaryote, like a bacterium; if your cells have nuclei, you're a eukaryote, like me. The first eukaryotes show up in the fossil record about 1.5 billion years ago.

Finally, between 800 million and 540 million years ago, life took its third big step: complex plants and animals arrived on Planet Ocean. Those hard-working, oxygen-producing, genetically facile eukaryotes diversified to become plants, invertebrate animals, and, in the Cambrian, the first hint of chordates. Probably. The deep mysteries of the process abound. We know for sure from the fossil record that taking advantage of a changing ecology, seeking nourishment, avoiding being eaten, and having some fantastic luck keeps the show on the road. The first multicellular creatures were probably colonies of single-celled organisms that adapted to a division of labor, coordination, and interdependence. All that complexity had slowly worked its way into the cast on Planet Ocean by the start of the Cambrian, 540 million years ago. What followed was an outburst of diversity and evolutionary horsepower more fertile than any before or since. We call it an explosion, but it was really more like the beginning of a fugue that would invent itself over and over from the debris of living and dying to inspire wonder, peace, and sorrow among those of us who are aware of our own existence. The clearest notes of complex life's first songs echo in the dark shale of the Canadian Rockies. We heard them just after the turn of the century, though we didn't hear the tune clearly until the mid-1970s.

But what's a few decades among eukaryotes?

be anaerobic—deadly allergic to oxygen. Quickly, oxygen was boss, and most anaerobic cells were the victims of a mass extinction that cleared the field for further evolution.

The leap from inorganic ditch water to oxygen-making organic scum was the first third of the miracle of life. The second was the development of cells that have nuclei and contain a coded substance we now call deoxyribonucleic acid, or DNA. Probably because the new oxygen atmosphere shielded Planet Ocean from ultraviolet radiation, allowed

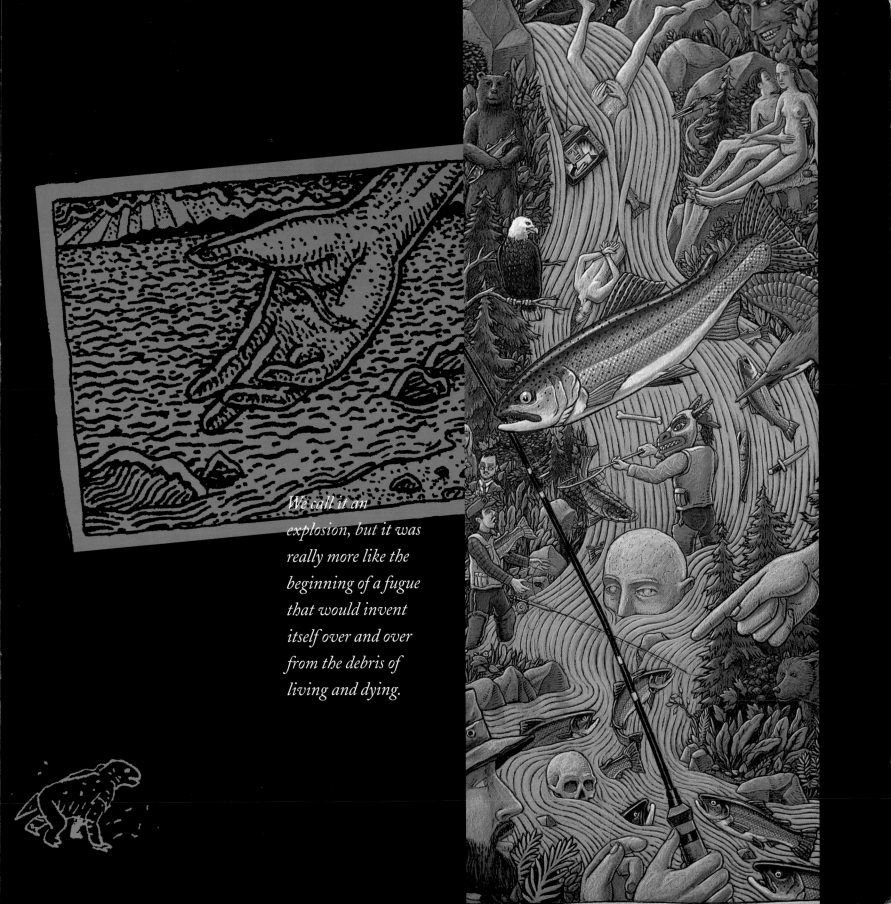

We call it an explosion, but it was really more like the beginning of a fugue that would invent itself over and over from the debris of living and dying.

As Brad and I began the adventure of Planet Ocean neither one of us knew quite what direction we were headed in. We decided to trust in the collaborative process and to take our creative cues wherever we got them, following instincts, personal interests and always looking for the fishy segue.

The Gulf War was raging at the time we started - strange surreal images beamed into the living room - WAR - eternal war. I had such strongly conflicting emotions about the whole thing - the moral barometer spun wildly.

At the same time we were both plunging ourselves into all things Paleontological. Stephen Gould's book Wonderful Life when fully digested, had a profoundly unsettling message - we are here by chance - pure luck of the big Cosmic draw.

Complex life had burst full force onto the planet in the Cambrian - astounding in its diversity and seemingly resilient. A floodgate let loose, mutating over the millenia until we happened along - wayward Dipidistian fish.

Studying the moldy fossil record - FROM LOBE FINS TO HIGH TOPS - one immediately realizes that extinction is a common occurrence. It's a short leap from there to the understanding that one of the biggest MASS extinctions to hit the home planet is happening right now - at our hands.

It was enough to blow anyone's mind. Brad and I conversed almost daily like seventh graders hooked on a wild science project. Finally we took the great road trip across the West to the fossil beds.

Inexorably drawn on ...

SPLENDOR 'NEATH THE TRILOBITE

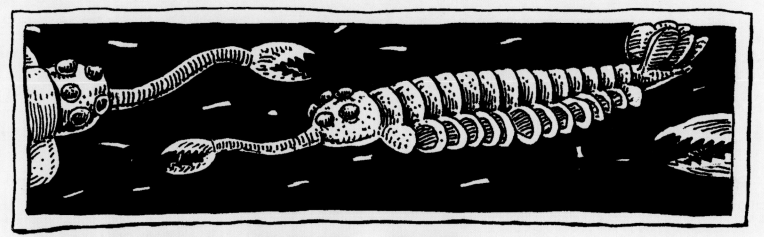

READY, SET, GO

STONE BUGS, TRILOBITES, AND OCEANS IN THE SKY

THE GANDY DANCERS called trilobites stone bugs, and they were as common as flies when the railroad builders chipped and blasted away the swirling layers of the Canadian Rockies. At the end of the nineteenth century, these hard workers were busy pushing the promises of the industrial age through the mountains to the Pacific, so their curiosity about the odd rocks with eyes and legs was momentary at best. But the trains that would eventually haul timber, ore, gold, fish, and wheat from the wide-open spaces west of the Rockies also, incidentally, hauled fossil hunters to the fossils.

Among them was Dr. Charles Doolittle Walcott, the fourth secretary of the Smithsonian Institution, an expert on the stone bugs and a gentleman who carried the Victorian mission of discovery like a battle flag. (He was also the director of the U.S. Geological Survey, president of both the National Academy of Sciences and the American Association for the Advancement of Science, and custodian of American forest reserves.) Walcott's field specialties included fossil arthropods, trilobites in particular, and in the wildly folded sediments of the Rockies, he and others found the most ancient of specimens.

Canada's celebrated mountains are nothing less than oceans in the sky, exposed when the continental crust buckled and rose 110 million years ago, since eroded by wind and water to their present shapes and sizes. Rolling waves of rock formed as layers of the sea bottom, carrying the evidence of life in the Cambrian and early Ordovician ages, when everything alive lived in the sea. Trilobites thrived for about 300 million years and were the most diverse and successful animals on Planet Ocean until the Permian extinction claimed the last of them—along with 96 percent of all life in the sea—225 million years ago.

TRY A LIL' BITE

O' TRILOBITE

Trilobite fossils are found on every continent, often secondhand in archaeological digs, which means early humans were attracted to the stone bugs, too. Walcott and others were fascinated by trilobites because of the length of their ride on Planet Ocean, their immense diversity, and their astounding, complex body plan. Some creationists say that the tons and tons of these messages from the past are really the discarded castings of critters that didn't make it past God's quality-control checkpoint. Hardly discards, these hardy travelers were segmented into three lobes—head, body, and tail, hence the name tri-lob-ite—contained in a jointed exterior skeleton. This hard shell of calcite was biochemically mined from the sea, and a trilobite grew and shed many shells in a lifetime. They inherited and refined a primitive circulatory system consisting of a heart and vascular ducting, a respiratory system with gills, and a highly developed nervous system hooked into something like a brain. And then, too, trilobites were dealt an evolutionary pat hand that gave them top critter status: all but a few had eyes, the first ever. If trilobites could see while all their fellow invertebrates were playing pin-the-tail-on-the-donkey wearing blindfolds, it's not hard to figure out who was going to eat whom.

The name of the trilobite phylum, Arthropoda, means "joint-footed" and includes such surviving residents of Planet Ocean as insects, scorpions, spiders, shrimps, crabs, cockroaches, and lobsters. The oldest, most primitive trilobites scuttled into the fossil record with no apparent transitional form because their ancestors probably had soft, nonfossilizing bodies, which almost always disappeared completely. The fossils of trilobites most recently alive were collected on the high slopes of Mt. Everest, of all places, from 230-million-year-old rock laid down just before the Permian extinction killed off the final members of the clan. Among the last to go were the proetids, the most common of the ten thousand species of trilobites to have lived over the ages, a very unspecialized gang. They would eat anything and breed anywhere, and they made themselves as unattractive to predators as possible. We all have relatives like them. From proetids and their success and longevity, an evolutionary rule of thumb has emerged: "The more specialized a species, the less able to cope with change it will be once the inevitable happens and old habitats change beyond the point of recognition," as the inspiring paleontologist Niles Eldredge put it in his book *Life Pulse.* In other words, generalists usually outlast specialists, and evolutionary progress is not necessarily a matter of refinement.

The lesson of the proetids solves one of Darwin's dilemmas, too. Why, the great naturalist wondered, if survival of the fittest yields slow, steady progress over time, are there any primitive animals around at all? Why weren't prokaryotic bacteria completely replaced by eukaryotes? Why haven't all the primitive plants and animals that still exist continued to evolve, becoming complex roses or steelhead fishermen? Well, if sheer survival over time is the product of good genes and good luck, the simpler species that make it through millions of years have no reason to evolve. Ninety percent of success is just showing up. Ask an arthropod, like a trilobite or a cockroach. As a fossil or a household pest, it will tell you that diversity and cooperation as much as competition and specialization, ensure survival. Generalism won't get you to Carnegie Hall with your cello, but a cockroach doesn't need a cello.

TRILOBITES

Trilobites were among the first living beings with eyes. These hardy arthropods popped into the fossil record in the Cambrian, about 510 million years ago, and were successful predators through the Ordovician, before any life had come ashore. Over 10,000 species of trilobites rode Planet Ocean for about 350 million years until the great Permian extinction claimed the last of them.

SOFT BODIES, HARD PARTS

CHARLES WALCOTT was fifty-nine years old in the summer of 1909 when he went trilobite hunting in the Rocky Mountains with his wife, Helena, his children, Helen, Stuart, and Sidney, and his scientific assistant. Family field trips are the rule among paleontologists, and the Walcotts—like the Sternbergs, Bonners, Goulds, and other paleo-families—often spent their summers in the fossil beds where spouses and children made major discoveries themselves. In one account, now believed to be campfire legend rather than truth, Helena Walcott's stumbling horse led the entire world off the trail to a slab of rock that contained the first of the fantastic fossils of the Burgess shale.

Maybe it was the luck of Helena's foul-footed mount, maybe not, but in late August 1909, in the dark talus of a ridge between Mt. Burgess and Mt. Wapta, Charles Walcott found not only some new notes in the fossil record but an entire symphony. A fossil is any evidence of ancient life—bones, teeth, shells, tissue, tracks, or traces—that has been buried rapidly, insulated from oxygen and decay-producing organisms, and has remained buried and undisturbed in sediment that became rock. Trilobites were readily preserved because their hard, calciferous shells prevented them from being destroyed by the heat and pressure of rock making. Not so the soft-bodied invertebrates, denizens of Planet Ocean for millions of years before trilobites figured out hard bodies. Though Planet Ocean was alive with soft-bodied creatures for maybe a half-billion years before trilobites, not many made it into life's chorus in the fossil record.

In the Burgess shale, Walcott found the fossils of soft-bodied animals, some with a few hard parts,

bearing testament to the glories of an age of unlikely diversity. The sudden collapse of the face of a massive algae reef and perfect conditions on the sea bottom gave us these shadowy remains in what would become dark Cambrian rock. Walcott's fossils still keep squads of paleontologists busy trying to figure out who's related to whom. The Burgess shale became an instant legend, more enlightening even than the Niobrara chalk, Green River shale, Solnhofen limestone, and the Morrison formation, each of which spun the paleontological compass when their rich fossils were discovered.

Walcott reported that the fossil animals of the Burgess shale were almost all arthropods—and made the biggest mistake of his distinguished career. He was an arthropod specialist, inclined to think the proto-trilobite, or prelude animal to arthropods, was surely somewhere in the record of the sea. He was after it, and, more critically, he and everybody else studying rocks and fossils during his era of dominance were lashed to the Darwinian wheel of steady progress. It was unthinkable that primitive organisms would not fit somewhere in the phyla of their more complex successors. Walcott's fossils were a sensation at the time, adding more than 80,000 nearly perfect soft-bodied specimens to the collection at his beloved Smithsonian Institution and drawing visitors and requests for loans from around the world. The real surprise, though, would come almost seventy years later when Harry Wittington took a hard look at the fossils of the Burgess shale, unburdened by the trappings of nineteenth-century anthropocentrism and certainty.

Fossils can be hunted in the field, where you also get to swat flies, eat bad food, sweat, and sleep on hard ground. But many of the greatest discoveries have been made in cabinets in cool museum

So·Who·Can·Explain·it
tribolite·bite
CAMBRIAN
DIGGING·ALL·THINGS
g room·to·that·trilobites·paleo·love·thing
That·ancient·love·of·arthopods

**That Ancient Love
of Arthropods**

basements where the fossils have rested, out of sight, since they were collected from the rocks. They are all cataloged and tagged with cards that list the place of origin, period, formation, name of the fossil hunter, and tentative identification. Only the most interesting, though, are described and published to begin that wonderful, slow dialogue of scientists in their journals, as were many of the Burgess shale specimens. In the cabinets and trays of several museums and universities is where Wittington, a Cambridge invertebrate paleontologist, came across the Burgess shale fossils. He then spent twenty years studying them.

Wittington had migrated to Cambridge from Harvard, and in both groves of academe he absorbed the messages in the shadows of these odd animals. In 1971, he finally concluded that Walcott had blown the job. Yes, some of these creatures were indeed arthropods. Many, however, were members of absolutely no known phyla, with body plans that fit none of the twenty-six living animal types. The Burgess shale contains, Wittington said, not only extinct species of ultimately successful types of animals but entire forms of life that departed Planet Ocean forever, even as the Cambrian was exploding. The feast of diversity, it seems, was even more sumptuous than Walcott ever thought possible, and survival—that comfortable Darwinian conclusion—might be a matter of pure luck.

Stephen Jay Gould, a great paleontologist and popular natural history commentator, became the third éminence grise to review the implications of the Burgess shale fauna for life as we know it. Not coincidentally, he replaced Wittington at Harvard. In his book *Wonderful Life*, Gould defends Wittington, but he goes a step farther into grand theory and asserts that sheer luck has enormous consequences

Yoho British Columbia. Perhaps the most
...ful stretch of highway on our planet. A

... the good things in life. ... "road moment" ... cruisin ...

in the history of life. Had a different collection of these delicate animals survived mud slides and mass extinctions, Gould says, human life simply would not have evolved.

Then came the inevitable outcry: *"But we thought humans were a sure thing in the evolutionary conga line."*

Gould *is* pretty hard on Walcott, obviously respectful but more than a little miffed at the theologically bound, Victorian tyranny that corrupted scientific inquiry during Walcott's career and well into his own. Gould says that although Walcott worked on the Burgess fossils with no technological or scientific limitations on his vision, his conclusions missed the mark because of social and religious folly. Neither Walcott nor Darwin, only fifty years his predecessor, could come to terms with life on Planet Ocean as anything but a procession of understandable stages. But Gould is quite willing to roll around in the ultimately joyful terrain of an uncertain universe, restraining himself from shoehorning life into patterns that suggest meaning and purpose and allowing life on Planet Ocean to describe itself to us as it will. A couple of years after *Wonderful Life*, of course, critics are taking aim at Gould and successfully showing that a lot of these Burgess shale critters are arthropods. But not all of them.

Though acrimony is often a side dish at the banquet of scientific conflict, the entrée is discovery. If we all agreed all the time, we'd never learn anything, a notion that prompted Thomas Huxley, a wizard of nineteenth-century inquiry, to comment, "In science, the next best thing to being right is being wrong."

PILGRIMS AT THE BURGESS SHALE

LIKE CHARLES WALCOTT, Ray and I made our way to the Burgess shale through the railroad town of Field, British Columbia, kidding that we were finally doing some fieldwork. We were first-date nervous. We were also as excited as lottery winners on the way to pick up the check. In early May, the Canadian Rockies deliver an invigorating repertoire of snow showers, bright sun, chilly mist, and vagrant light that toys with the rough shapes of the mountains against chrome-blue sky. We stopped a few times on the shoulder of the two-lane blacktop, risking thundering semis and swivel-necking tourists for a glimpse of the ocean floor that is now the Canadian Rockies. The peaks and ridges looked like actual waves, with crests and all the drama of giant surf.

We were anxious to get to Field but couldn't resist taking a roadside break to watch a mile-long freight train enter and leave the amazing tunnel the railroad had built to slow down trains that descend 4,000 feet from the Great Divide to the Kicking Horse River valley. At the top end, the train enters and, invisibly, goes into a tight, spiraling, 360-degree turn, actually running under itself. Then, still in the tunnel, it heads straight down a gentle grade and into another 360, losing elevation without building up deadly speed. Before the spiral tunnel, when Walcott and his family rode these rails into this country, speeding trains routinely crashed at the bottom of the long, steep grade.

We stood there on the wooden platform overlooking the engineering spectacular and watched busloads of sightseers point and take pictures of each other pointing at the trick train. From our perch,

Charles Walcott was fifty-nine years old in the summer of 1909 when he went trilobite hunting in the Rocky Mountains.

OVERLEAF: Critters of the Burgess Shale

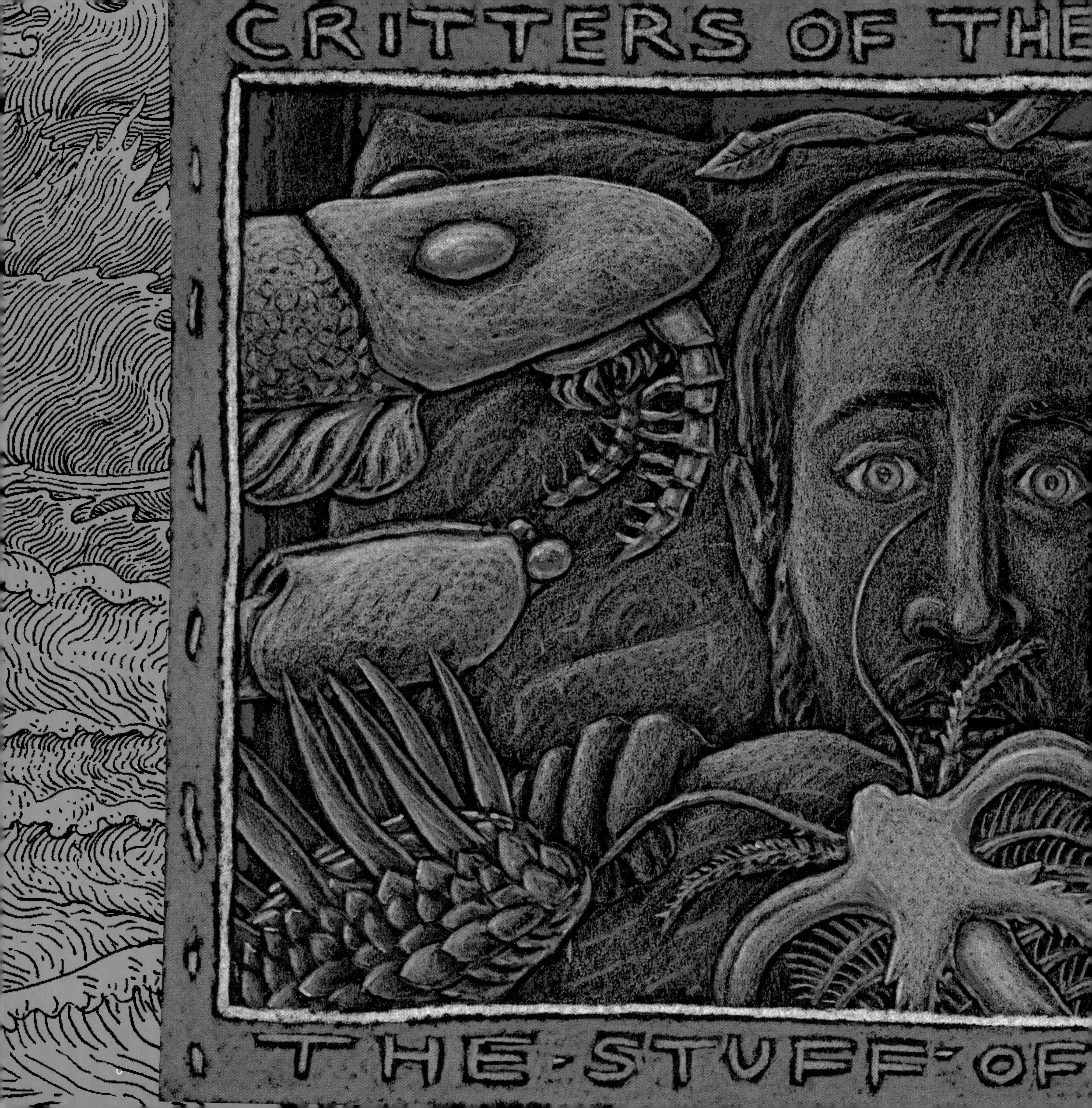
CRITTERS OF THE
THE·STUFF·OF

we could hear the screeching brakes and clashing drawheads of the freight in the thin mountain air, sounds as normal now as those of the runoff waterfalls and birds that Walcott would have heard eighty-three years before.

On topic, as always, Ray and I talked about assumptions, wild guesses, and paleontology, about how much of the fossil record was missing or fragmentary. And we marveled at the stupendous consequences of a discovery like the Burgess shale. From the spiral tunnel overlook, we could have seen Walcott's quarry with X-ray vision, had we known to look west through the mass of Mt. Ogden, across the Yoho Valley, through Mt. Field, and onto the ridge where he found his fame. Instead, we got back into the car and finished the drive into town. Without the benefit of a spiral tunnel, we hurtled down the Trans-Canada Highway with the Boss, loud on the tape player, reminding us that we were "Born to Run" as we flashed in and out of the mist like the mountains that were once the ocean.

High speed alone could have accounted for our pilgrim's nerves, but we were also about to get as close as we could to the Burgess shale after months of reading and thinking about it. This was not just a visit to another roadside attraction; no Grand Canyon, Niagara Falls, or spiral tunnel. What had happened on the ridge between Mt. Field and Mt. Burgess in the summer of 1909 is much more akin to the discovery of the cosmic background radiation that gave us an entirely new story of our universe, galaxy, solar system, planet, and neighborhood.

In 1965, a couple of radio astronomers, Robert Wilson and Arno Penzias, heard static that they thought was caused by pigeons scratching in their antennae but that turned out to be leftover radiation from the fireball that created the universe. Their discovery, the first evidence of the Big Bang, confirmed suspicions arising from Vesto Silpher's realization in 1913 that the galaxies are moving away from each other, that the universe is expanding, and that there are billions and billions of galaxies, each with billions and billions of stars within them. If you put those news flashes together, you get the underpinnings for a scientific account of genesis that involves lots of time—and lots of uncertainty.

In the dark talus of Walcott's quarry, an earlier tale of uncertainty unfolded, a story that revised the notion that we can know enough about the present to fathom accurately what happened in the past. Though paleontologists, biologists, and other seekers have known for years that the Burgess shale fossils triggered a revolution, civilians like Ray and me had been shaken suddenly into the new paradigm. Luck, chance, fortune, and fate radiate from that wild ridge in the Rockies, at least as consequential in the flow of life on Planet Ocean as good genes, good works, or favored critter status.

The animals that form the Burgess shale lived before fish or any other chordate, so our most ancient relative might be among them. But most of the fossil animals were members of distinct groups of life, many of them vanished soon after the Cambrian, so our grand, grand-fathermother survived by winning the biological lottery, not by being any more fit than any of its reef-mates. Quite simply, if you went back in time to a day when all the Burgess animals were alive and started evolution all over from there, the odds of ending up with Ray or me or you are just about zero.

Cosmic uncertainty can be a touch unsettling to a couple of ex-Catholic white guys careening down a mountain road in the late twentieth century, but a roll-the-dice planet was utterly unfathomable to Walcott. He would have had to reinvent morality, ethics, and progress without divine payoffs for being good, pious, or a survivor of a mass extinction. The Burgess shale throws a serious wrinkle in life as we know it.

AT THE VISITOR CENTER

EVEN AS WE WERE blasting ourselves into new orbits of self-awareness on the way into Field, we realized the chances of actually getting up to Walcott's quarry were slim. First of all, the snow cover is ear-deep on the ridge until late June. Then, too, the quarry is far too precious to allow a couple of rubes

ANOMALOCARIS

The giant of the Cambrian, up to forty inches long, *Anomalocaris* brought terror to the sea. Almost every other critter in the fossil record from the time was mere inches or less in length, so Charles Walcott thought a piece of a single *Anomalocaris* was more than one animal. The body plan of this great swimming predator vanished after the Cambrian.

PRE-CAMBRIAN MEMORY

from Alaska to just walk up there and prowl around. We decided we'd get as close as we could, though, starting at the headquarters of Yoho National Park, home to the Burgess shale, or Schistes de Burgess as they say in bilingual Canada.

After the buildup of our morning's conversation and the exhilaration of being within a few miles of Walcott's grand discovery, we were surprised to find the parking lot at the visitor center almost empty. Where was everybody? Why weren't long lines of pilgrims shuffling around the low stone buildings? It seemed to us there should have been crowds, even on a weekday in early May, instead of three cars on a half-acre of wet, glistening asphalt with the peekaboo sun winking at us from the puddles.

"Kind of quiet," Ray said. But, mercifully, the doors were open, and we dodged what would have been crushing disappointment. Under a beamed ceiling and across an institutional linoleum floor, we faced a wall of glass exhibit cases, the arc of a service counter with a milling cadre of park rangers, signs for rest rooms in English and French, water fountains, a T-shirt and souvenir booth, several kiosks with television monitors, and another wall of photographic murals and drawings.

Just about everybody has been in a park visitor center. They're all over the place. In the twentieth century we have concluded that nature requires protection from us, so we have set aside hunks of nature and preserved them, a unique, civilized piece of business under the circumstances. Though trilobites were infinitely more successful and diverse than *Homo sapiens sapiens*, no species has threatened more destruction than we have in just a few million years. And no other band of humans in the history of Planet Ocean has spread its values and culture around the world as we, the gang the

Romans called barbarians, have in a few hundred years. Realizing how much we are capable of destroying, we have impounded entire regions as parks and wilderness areas. But the worst outcome of such caution is the disturbing sense that we are not part of nature ourselves.

Yoho Park happens to be in Canada, an early subscriber to the remarkable American invention of grand public preserves, a hundred-year-old idea on the day Ray and I strolled over to the display cases of 540-million-year-old specimens from the Burgess shale. Like a couple of kids who collect baseball cards walking through the turnstiles to their first real game, we stood there ooohing and aaahing, pointing and naming in the stage whispers commanded by powerful icons. We had seen hundreds of pictures, but before us at that moment were five-eyed *Opabinia*; *Wiwaxia*, looking like a punk trilobite; *Hallucigenia*, the nightmare walkingstick; and the elegant *Marrella splendens*, the arthropod lace crab and the insignia on the flag of the park, an ancient beauty very common in the Burgess shale but found nowhere else, so far.

Most of the twenty-eight Burgess animals were there in the case, either the real things in actual pieces of the dark, distinctive shale or casts made of the ones too valuable to set aside for a visitor center. Public viewing is nice, but most of the rarest and best fossils circulate only among paleontologists for research. Just their names were an anthem to us, though, as we gazed through the glass and muttered the litany of genera: *Choia, Naraoia, Nisusua, Anomalocaris, Leanchoilia, Pirania, Scenella, Burgessochaeta, Ottoia, Burgessia, Olenoides, Vauxia, Molaria, Hallucigenia, Sidneyia, Hyolites, Louisella, Dinomischus, Canadaspis, Eldonia, Waptia, Aysheaia,* and *Pikaia gracilens*. Except for the two-foot giant *Anomalocaris,*

Cosmic uncertainty can be a touch unsettling to a couple of ex-Catholic white guys careening down a mountain road in the late twentieth century.

most of them were tiny, some almost microscopic, and, without labels, hard to identify from the drawings we'd seen.

One of the lucky strokes of the Burgess extinction event was the speed with which the killing avalanche happened, so the fossils depict many profiles of the animals frozen in a tumbling rush toward their deaths. To untrained eyes like ours, the results resembled stains on a bug-spattered windshield. But among these midget roadkills are the ancestors of most of the modern creatures of Planet Ocean and, most remarkably, many entire tribes that didn't make the cut—like *Opabinia regalis*, a clown of an animal with a segmented trunk, five eyes on stalks on a distinct head, and, clearly in the specimen we looked at, gills on its body. *Opabinia* was one of Walcott's would-be arthropods, later reclassified in its own

To untrained eyes like ours, the results resembled stains on a bug-spattered windshield.

phylum by Harry Wittington and a classic example of the significance of the Burgess shale fauna. It is extremely rare; only ten have ever been found, nine of those by Walcott himself, so Ray and I were probably looking at a replica.

Because the ancient avalanche preserved many of the Burgess animals in three-dimensional relief, they can actually be dissected using microscopes, mini-drills, and selective dissolving chemicals. *Opabinia* was the turning point for Wittington in his realization that Walcott was wrong, that all life does not proceed steadily to some conclusion predictable from the present and the past, that life on Planet Ocean has unwound with a lot of luck. "The Burgess shale," Wittington concluded in 1975 from his study of *Opabinia*, "contains other undescribed segmented animals of uncertain affinities." And as Wittington's interpreter, Stephen Jay Gould pointed out, "We are

awestruck by *Tyrannosaurus*; we marvel at the feathers of *Archaeopteryx*; we revel in every scrap of fossil human bone from Africa. But none of these has taught us anywhere near so much about the nature of evolution as a little two-inch Cambrian oddball invertebrate named *Opabinia*."

"Can we help you?"

Ray and I turned from our reveries at the fossil case to see a pair of park rangers smiling at us over the counter, obviously amused by our chatter and exhilaration on an otherwise slow day. In five minutes, they confirmed our suspicions: a hike to the Burgess shale itself was not in the cards. We could come back in July and sign up for a guided trek and maybe even meet Desmond Collins, the latest in the line of paleontologists working on the site. For the time, we had to content ourselves with the exhibits in the visitor center, a videotape or two, and a distant view.

The videotapes weren't bad. We watched Collins and his crew at work and heard that during the 1990 season they'd found new species from the Cambrian explosion in the Burgess shale and at sites in China, Greenland, Wyoming, and Colorado—including a kind of winged worm and a pincer squid. We sensed the wonder and delight in the voices and faces of Collins, Gould, and the other narrators whose journeys in time were adjusting forever the way we think about life's bloom on Planet Ocean. They seemed to reserve their most astounded tones for the profound discovery that evolution yielded maximum diversity right off the bat and not, as Darwin figured, increasing diversity through time. The implications of lottery and luck in such a process challenge the myths we humans have created to deal with the fears and mysteries of life and urge us to come to kindly terms with unvarnished uncertainty. The Burgess shale is a hell of a good argument for just lightening up.

So we couldn't get to the quarry. No problem. We'd go have a look from Emerald Lake, the best vantage point, the counter people told us, at the end of a short drive on a clear road. The parking lot at the lake was packed with cars when we got there, and we thought they must belong to pilgrims who knew where to go in the first place, bypassing the deserted visitor center in Field. We took a well-tended trail to the lakeshore but away from the main traffic pattern to and from a lodge in the woods which, it turned out, is the tourist attraction that filled the parking lot. Bad coffee, decent pie, and clean rest rooms draw a bigger crowd any day than a distant view of the scene of a cosmic bulletin.

Our path led us to a quiet dock and a curio shop, tended by a woman coincidentally named Helena. She was a chatty greeter, probably more at home in midsummer when her shop was in the tourist mainstream instead of in a backwater as it was on that day in the shank of winter. In the shop, Helena was dealing rental canoes, Emerald Lake T-shirts, film, batteries, candy, and sunblock—but not a single memento of the Burgess shale; no Burgess Burgers, no *Opabinia* masks with five eyes and a nozzle, no hunks of shale or posters of Walcott, Wittington, and Gould, no black velvet paintings of *Marrella*, *Anomalocaris*, or *Pikaia*, no *Aysheaia* keychains or *Hallucigenia* jackknives. Nothing.

"So, you ever heard of the Burgess shale?" Ray asked Helena. She said, "Oh, sure," and nodded in the general direction of the lake. "It's across the water and up to the right, but you can't go there." We walked out onto the dock anyway and scanned the rumpled terrain over a rack of forest green canoes and across the glassy water where flotillas of lovely grebes, loons, and mergansers burbled and scooted. The scoured flank of Mt. Wapta, we determined,

was a couple of miles away, up and to our left, a forbidding, crumbling slope of ancient rock that looked very fossily. But across a forested saddle to the right, the blocky butte of Mt. Field loomed, and beyond its raw mass, the peak of Mt. Burgess, named for a former governor-general and for many years the central image on the Canadian ten-dollar bill. There, on the ridge that joins Mt. Field and Mt. Burgess, Helena told us through the door of the shop, was where Walcott found his fossils and, eventually, a new version of the story of life. "It's kind of dead today," Helena said, "but a lot more people have been showing up lately to take a look. He found some pretty weird stuff up there, I guess."

On the day in the summer of 1909 when Walcott and his entourage climbed on horseback through Burgess Pass from the town of Field and out along that windy ridge, life was a much simpler proposition than for Ray, Helena, and me there on the dock at Emerald Lake in 1993. Then, the explanations of science brought comforting conclusions, rules, laws, and order. People felt safe building their lives on batting averages, a good job, and the promise of an afterlife. But like the much later revelations of plate tectonics in our version of the geologic story and background radiation in the origins of the universe, the remains of the Burgess animals delivered their ultimately unsettling message: Life is uncertain.

Walcott and most of his contemporaries maintained a firm belief in an orderly, hierarchical process of life's emergence on Planet Ocean, with humans in the central role and, despite their best intentions to the contrary, a grand, complex blueprint tacked to the wall. But complexity is not a matter of design, and we don't have to assume a designer to understand the origin of life. The model for evolution that is contained in an engineered result, like the spiral

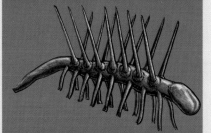

HALLUCIGENIA

Walcott thought *Hallucigenia* was a worm. Wittington thought it was another of the Cambrian's short-lived phyla. After new *Hallucigenia* specimens turned up in China, paleontologists realized that identifying it was a problem, in part because they had been looking at it upside down. Now, this weird creature is listed as an armored lobopod, an ancestor of modern, centipedelike velvet worms. But stay tuned.

WOMAN FED UP WITH TRILOBITES

Bad coffee, decent pie, and clean rest rooms draw a bigger crowd any day than a distant view of the scene of a cosmic bulletin.

tunnel for instance, is not fundamentally mysterious. We know that somebody drew the tunnel on a drafting table, that others took measurements in the field, that laborers blasted through the mountain with enough precision to follow the plan, and that eventually crashes at the bottom of the grade were stopped. We're most comfortable with a process that includes purpose and predictable outcome and tend to include it in our myths and explanations for the evolution of life. The railroad builders had a purpose and designed their tunnel with sufficient complexity to achieve that purpose.

A process by which life could have evolved without a designer was Darwin's big contribution to the scientific story. He left some holes that were filled when we finally figured out that the planet, in fact, is 4.5 billion years old. Others were filled when we stumbled across the diversity in the Cambrian, which strongly suggests that survival is as much a matter of luck as design. Still, if something as simple as a mountain tunnel needs a designer, it's hard to shake the suspicion that a human body doesn't. According to a reputable survey in 1992, more than half of all Americans disagree with the basic theory of evolution and think a designer, a.k.a. God, had a direct hand in the emergence of life on Planet Ocean. Another 30 percent think a designer

cleverly started the process and then let it run, a category of speculators that includes many scientists. Less than 20 percent think life as we know it is possible without a designer.

The powerful message of the Burgess shale is leaking out, though: No special biological favors or status have been bestowed on *Homo sapiens sapiens*, and no intent or purpose is apparent in the selection of animals or plants for survival or extinction. "Wind back the tape of life to the early days of the Burgess shale; let it play again from an identical starting point, and the chance becomes vanishingly small that anything like human intelligence would grace the replay," writes Stephen Jay Gould, like an eloquent ventriloquist's mannequin speaking for the fossils, who demand to be heard.

Considerable bickering among paleontologists continues over the bold classifications Wittington, Gould, and others made of the animals of the Burgess shale. In Cambrian rock in China, paleontologists have found a fossil enough like *Hallucigenia* to reclassify the weird Burgess animal as an onychophoran, a kind of fat worm found in modern tropical forests; *Wiwaxia*, others say, is directly related to modern marine worms called polychaetes; still others have reevaluated many Burgess animals back into the four modern arthropod phyla. Nobody, though, disputes the two central revelations of those miraculous fossils: The diversity of the Cambrian explosion is unparalleled, before or since; and luck is every bit the equal of natural selection on the way to *Homo sapiens sapiens*.

HEALING THE CHORDATE WITHIN YOU

A LOT OF PEOPLE have a hard time imagining early primates as our ancestors, let alone *Pikaia gracilens*. Walcott was sure it was a worm, another polychaete, and he named it after Mt. Pika in the Rockies. Not until one of Wittington's graduate students, Simon Conway Morris, painstakingly dissected and described these tiny denizens of the Cambrian ocean sixty years later did we get the message. *Pikaia* had a proto-spinal cord, the crucial, defining characteristic of our own phylum, and its survival began a chain of beings that now includes fish, birds, and mammals, the vertebrates. None of the other Cambrian fossils show any sign of being a candidate for first chordate. Between *Pikaia* and us, though, is at least one chordate that's not a full-fledged vertebrate, a tiny fishlike creature called a lancelet, or amphioxus, still in residence on Planet Ocean. Then come the first true vertebrates, the agnathans or jawless fish.

Most paleontologists believe other Cambrian animals also might have been equipped with a proto-spinal cord, but none have been found. Only about thirty specimens of *Pikaia* exist, and when you compare their relative populations to the thousands of fossils of Cambrian arthropods or the true worms, it's obvious we chordates were probably low in the pecking order of that period and were at grave risk.

"There's our hometown," I said to Ray as we stood gazing up at Walcott's quarry, feeling the chill of late afternoon rippling across the lake. On the drive back to Field through the waves of rock, I couldn't banish the vision of that ancient reef collapsing to entomb the unlucky crawlers, swimmers, and burrowers of the Cambrian sea. I guessed at the sound of that avalanche and the surface ripples lapping the barren, as-yet-lifeless shore in the aftermath. I tried to imagine what caused it—an earthquake, or an asteroid strike. The mass dying of the Burgess animals was probably local, but any one of many subsequent global extinctions over 540 million years could have removed Chordata from the picture and did remove relatives of other phyla that survived the ancient avalanche, like trilobites. If the descendants of *Opabinia*, for instance, had hung in there past the Cambrian and *Pikaia*, instead, had taken the big hit, the priests and scientists of our vertebrate tribe wouldn't be wondering how we got here, Helena wouldn't be selling trinkets at Emerald Lake, and I'd be typing this with my nozzle.

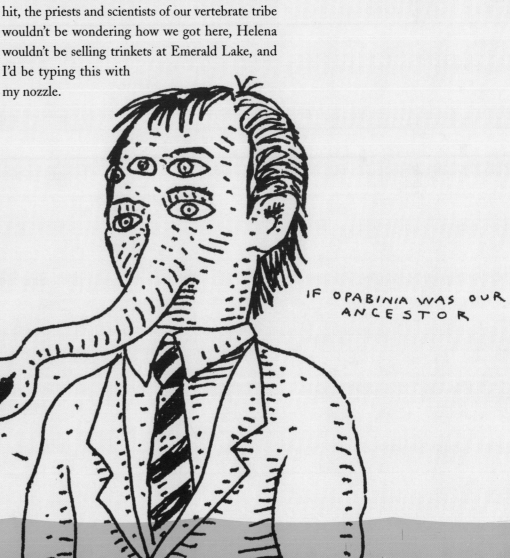

IF OPABINIA WAS OUR ANCESTOR

PIKAIA'S CHILDREN

WHAT DO WE TELL THE KIDS?

THE CHILDREN FLOWED into the Smithsonian rotunda like a river of howling energy from a steady procession of school buses. Maybe they were shrieking because they'd run smack into a life-sized African bush elephant that rose in the center of the entrance hall, or because their footfalls on the marble floor resonated like thunderclaps. Maybe it was just the exuberance of freedom after a long morning's bus ride from Richmond, or Baltimore, or any one of hundreds of towns whose teachers take their children to the Museum of Natural History at least once a year. The voices of the grown-ups with them wove in deeper tones as they herded their classes and families with promises of moon rocks, souvenirs, and, the main attraction, dinosaurs. Most of the kids were of prime dino-craze age, about six to twelve.

The galleries of the great museum sprout from

the rotunda on four tiers off a central hall and into a pair of opposing wings, containing just a small fraction of the 118 million specimens in the world's greatest cache of natural treasures. Perfectly prepared examples of most of Planet Ocean's birds, mammals, fish, insects, and other invertebrates occupy orderly aisles of cases, dioramas, and free-standing displays of the larger animals and fossils.

You can walk in off the street and, for free, stand within inches of rocks brought from the moon, the largest diamond ever found, or the 4.6-billion-year-old Murchison meteorite, a remnant of creation that surprised everybody with its cargo of amino acids, the stuff of life. You can pace off the 92-foot length of a perfect replica of a blue whale suspended from the ceiling, one of the largest animals that has ever lived. Or you can consider Martha, the passenger pigeon, now mounted on a branch in a case but once the last surviving member

Einstein was a fish. So were Mozart, Virginia Woolf, and Thomas Jefferson.

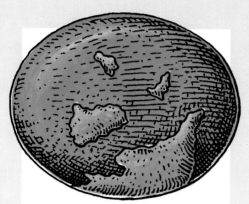

LAND AND SEA DURING THE DEVONIAN
365 to 410 million years BP

The first sharks, insects, and spiders showed up in the Devonian, and fish, our vertebrate ancestors, came ashore on the single landmass called the Red Sandstone Continent. The period ended in an extinction event that took out about 70 percent of all marine animals, but that seems to have left most of the new terranauts alone.

of her species until she died on September 1, 1914, in the Cincinnati Zoo. She is lean, not like the plump city pigeons with which we are familiar, and when you take a moment to examine her markings and imagine her in flight, she is beautiful in that way of wild birds. Passenger pigeons once darkened the skies over North America and, not incidentally, triggered a commercial orgy by market hunters, who dropped them by the thousands with bird-shot cannons. Martha's death was a front-page tragedy, and Americans wept. The specter of species extinction because of human greed had arrived like a relative with bad news in a late-night phone call.

Most of the Smithsonian's collection is in a huge warehouse in Maryland, shown to the public only in carefully staged exhibits after years, even decades, of preparation. The treasures are held in trust for the world, and the top banana of the Smithsonian is none other than the vice-president of the United States, who heads a committee of congressmen, scientists, and well-heeled benefactors. In the stories they chose for display, a secretary, such as Charles Walcott, and an enormous staff reveal our culture's most profound sensibilities about life. The stories change as our collective view of the universe, the world, and life on Planet Ocean is revised by discovery. More than anything, the Smithsonian and other natural history museums give our children what is often their first glimpse of the variety, elegance, and wonder of life apart from their immediate surroundings.

In 1990, the Smithsonian Institution opened "Life in the Ancient Seas," filling an entire wing with the story of evolution from 4 billion years ago to the day I walked through the riot in the rotunda and into a darkened alcove to begin the journey through time. The dim light and piped-in soughing

of ancient waters hushed even the most raucous children, and the lifeless silence of the early Archean began the story.

If you had a wristwatch calibrated to geologic time, its hands would move ultra-microscopically during your life. But they would measure the duration of the Archean with a sweep from high noon to ten o'clock. Contemplation of this most ancient, mysterious, and debatable of eons begins with the mental gymnastics of imagining 4 billion years, or 1 trillion 4 hundred billion daily rotations of Planet Ocean on its axis. Most likely, even the axis itself wobbled and shifted many times during the Archean, and the landmass and fragments moved around the uncertain planet like lily pads on a windy pond. The crust-building engines performed their chores, mountains rose and subsided, and under an atmospheric blanket only an eyelash thick on a planetary scale, the air and ocean began the quest for a deep breath.

During most of the Archean, oxygen made up only 1 percent of the air and water in free or dissolved form, far too little to sustain life-as-we-know-it. By the Cambrian, 4 billion years into the story, the oxygen level in the atmosphere had reached 7 or 8 percent, as compared with today's 20 percent. Just enough to support trilobites, but a bush elephant or a nine-year-old human wouldn't have had a chance. During the Archean silence, the headline act on Planet Ocean was performed by the great biotic virtuoso, photosynthesis, a cast of zillions of single-celled bacteria, and eventually a slightly improved character, blue-green algae, a.k.a. cyanobacteria, which has a distinct nucleus. The next time you cut down a tree or read a newspaper, remember that making oxygen, not Fords, is job one.

Above the cluster of suddenly restless children with me in the Archean, a spooky mural depicted a coastal scene, with the odd light of a too-white sun flashing through high, thin clouds. A pair of volcanoes that would be more in place on the moon dominated the desolate scape over a sea that looked hot and somehow thicker. And where the water scribed land's edge, clusters of lozenge-shaped mounds guarded a rocky beach. These weird mounds are the objects of prime interest here at the beginning of the story of life, because a couple of billion years into the Archean, blue-green algae started dating. Nothing heavy, a movie once in a while, and decaf coffee and pie afterward, but no reproduction.

Maybe they were cooperating in their photosynthetic chores by retaining heat or lowering the cooling effect of water by reducing their surface area, or trying to stay warm. It's hard to say. Something symbiotic occurred, though, probably the first instance of the cooperation that recent research has proven to be more common in nature than the competitive instinct. "Survival of the fittest," the phrase that rings through Western culture as the crystal of Darwin's theory of evolution, is only part of the story. Yet we insist that societies, businesses, personal relationships, and sporting events fueled by the competitive instinct deliver the strongest shot of nature's basic juice.

In those lozenges on the Archean beach, called stromatolites, is also the first evidence of multicellular life in the fossil record beyond faintest specks of organic debris. The mounds of thin layers of blue-green algae and calcium-rich sediment accumulated because the algae was encouraged to sprout filaments into the sediment layers and bond into these masses, some of which grow several feet high,

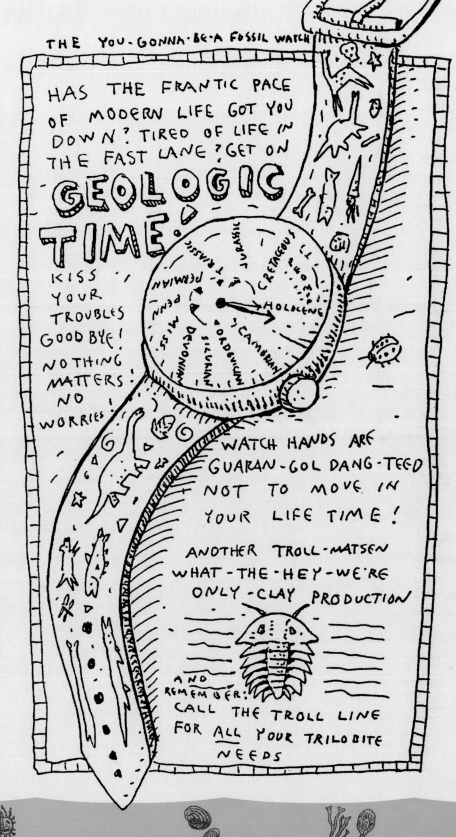

wide, and deep. Stromatolites still exist, most notably in the hot shallows of Shark Bay in northern Australia, where they have dodged predation by living in water too salty for the marine animals that would normally graze on them.

As fossils, stromatolites deliver the news that the genetic process began 3.5 billion years ago in the Archean, when a clearly defined nucleus evolved in the tribe of single-celled organisms. Quite possibly because of the interactions of single-celled organisms in colonies, cells got more sophisticated about making new cells. It was the first Junior Prom. Until then, reproduction was a simple matter of splitting a cell—the whole body of the animal. But with a nucleus, the reproductive duties could be separated from the economic functions of converting food to energy and maintaining a body. That was a dramatic, though not particularly lyrical, turn of events. We didn't completely figure out what had happened until the middle of the twentieth century when James Crick and Francis Watson unraveled the deoxyribonucleic acid molecule, the DNA double helix. Before humans solved the DNA puzzle, the endless repetitions of the genetic song since the Archean were purely spontaneous, producing plants and animals that have survived eons and others that hit dead ends in a few million years or less. Now, we not only know why babies look like other babies but we know how to engineer them, too.

Children with longer attention spans couldn't hurt, I thought, giving into cynicism as I wove my way through a knot of them clustered around a television set mounted in a kiosk. The TV comforted them, as its screen filled with an abstract,

repeating cartoon of the eukaryotic breakthrough. But the excitement of the legendary Ediacaran fauna was right around the corner. While the algae reefs and stromatolites are hot stuff, they were still just colonies of single-celled creatures so simple that any distinction between plant and animal was lost on them. Then, we know from the Ediacaran discovery, the eukaryotes made their move.

The kids shuffled along, looking for more action. Only the most reflective paused to peer at the replicas of what were among the first complex creatures to inhabit this world. Their filmy shapes appeared as hovering shadows, mirroring the current over the rippled ocean floor as they undulated like a great undersea forest of ferns. Some looked like leaves with stems attached to the bottom, some like fragile, transparent jellies with dangling lace cilia, some like hovering flatworms. The plants were distinct from the animals, a major breakthrough in itself, though nothing remotely resembling a vertebrate was to be seen in this Precambrian assemblage. *Pikaia* would not put in an appearance for a few million years.

The fossils on which this diorama is based are right up there with the great treasures of paleontology, equal in significance to those of the Burgess shale. First found in the Ediacara Hills of southern Australia by the geologist R.C. Sprigg in the late 1940s, but since revealed in rocks around the world, these are the lone whispers of Precambrian plants and animals, unlike anything found just a few million years later in the Cambrian. They are regarded as either rapidly evolving forerunners of the creatures of the Cambrian explosion or an evolutionary dead end. Like so many of the Burgess shale animals, they are all soft-bodied, and their survival as fossils for 570 million years is another paleontological

RHIPIDISTIAN REALIZATION

SUDDENLY THE CHILDREN UNDERSTOOD - THEIR ARMS WERE INDEED LOBE FINS

I wanted the children to come back, to stand there with me while I told them we were all from the same evolutionary hometown.

DUNKLEOSTEUS - A DEVONIAN FISH AS BIG AS A SCHOOL BUS

miracle. Very few resemble the later Burgess animals at all. But *Tribrachidium* could be a proto-echinoderm, kin to starfish, and *Spriggina* was a wormlike animal with a segmented body that might have been related to the eventually dominant trilobites.

The kids who swirled around my legs lasted barely a minute in Precambrian Ediacara, the last few million years of the Archean. They hurried across the hall without a pause in the Cambrian to look at the fossils from the Burgess shale and the diorama depicting those amazing animals in life. I stopped to linger and savor the actual fossils of my friends *Marrella*, *Wiwaxia*, *Opabinia*, and the rest, shipped in crates from Field, B.C., to Washington, D.C., in 1909 by Charles Walcott himself. *Pikaia*'s spot in the display was filled by a card stating that the fossil had been removed for study, as had *Hallucigenia*, no doubt because they are among the most sought-after messengers from the past since Harry Wittington turned up the heat on the Cambrian explosion.

I wanted the children to come back, to stand there with me while I told them we were all from the same evolutionary hometown and amazed them with the story of life in these dark rocks from a

EUSTHENOPTERON

Mother. *Eusthenopterons* were probably the first backboned terranauts to leave the sea for the future glories of cheeseburgers, fast cars, and television. Most paleontologists think this hardy lobe-fin fish became the first tetrapod, the ancestor of land-dwelling vertebrates, including us. Its skull and fin-leg bones are almost identical to those in fossils of the earliest amphibians, which show up in the fossil record in the late Devonian.

DEVONIAN D-DAY

1. **ICHTHYOSTEGA** *Late Devonian*
2. **CLIMATIUS** *Late Silurian*
3. **OSTEOLEPIS** *Devonian*
4. **PANDERICHTHYS** *Late Devonian*
5. **DIPTERUS** *Devonian*
6. **EUSTHENOPTERON** *Upper Devonian*
7. **RHIZODOPSIS** *Devonian*
8. **BIRKENIA** *Late Silurian*
9. **BOTHRIOLEPIS** *Upper Devonian*
10. **LONGANIA** *Upper Silurian*
11. **LUNASPIS** *Lower Devonian*
12. **AMPHIOXUS** *Cambrian to present*
13. **PIKAIA** *Cambrian*
14. **BOREASPIS** *Early Devonian*
15. **PTERASPIS** *Early Devonian*
16. **DREPANASPIS** *Early Devonian*
17. **CLADOSELACHE** *Late Devonian*
18. **ANGLASPIS** *Early Devonian*
19. **XENACANTHUS** *Late Devonian*
20. **PAREXUS** *Devonian*
21. **PALEOSPONDYLUS** *Mid-Devonian*
22. **LASANIUS** *Upper Silurian*
23. **ENDEIOLEPIS** *Late Devonian*
24. **HEMICYCLASPIS** *Early Devonian*
25. **PTERICHTHYODES** *Devonian*
26. **PTERASPIS** *Early Devonian*
27. **DUNKLEOSTEUS** *Late Devonian*
28. **GRIPHOGNATHUS** *Upper Devonian*

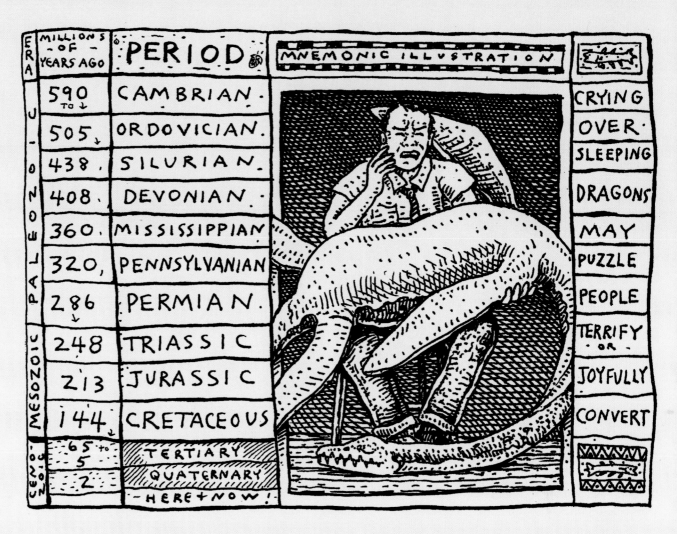

ERA	MILLIONS OF YEARS AGO	PERIOD	MNEMONIC ILLUSTRATION	
PALEOZOIC	590 TO ↓	CAMBRIAN.		CRYING
	505 ↓	ORDOVICIAN.		OVER
	438 .	SILURIAN.		SLEEPING
	408 .	DEVONIAN.		DRAGONS
	360 .	MISSISSIPPIAN		MAY
	320,	PENNSYLVANIAN		PUZZLE
	286 ↓	PERMIAN.		PEOPLE
MESOZOIC	248	TRIASSIC		TERRIFY - OR -
	213	JURASSIC		JOYFULLY
	144 ↓	CRETACEOUS		CONVERT
CENOZOIC	65 to 5 / 2	TERTIARY / QUATERNARY -HERE+NOW!		

Until the middle of the nineteenth century, the age of earth was a religious, social, and political number, instead of a physical truth.

mountain in Canada. And then I remembered how difficult that would be. How would I tell them the story of life on Planet Ocean is freighted with mystery, uncertainty, and wrong turns and that science, unlike preachers, priests, and shamans, is selling discovery instead of the comfort of knowing?

The time has never been better, though, for delivering the straight stories to our children. The great celebrations of discovery like those at the Smithsonian, themselves once burdened by creationism and the comforts of myth and certainty, are finally tacking away from that calm water in favor of the howling gale of uncertainty and wonder. When Walcott brought the fossils of the Burgess shale back to Washington, he said they fit right into the picture of life-as-we-know-it. For generations, we visualized evolution as an ugly primate rising in stages from all fours into the graceful carriage of an upright human. A lot of kids think cavemen and dinosaurs lived at the same time, and not too long ago at that.

WHAT TIME IS IT, ANYWAY?

THE TOWER OF TIME at the Smithsonian is a 27-foot vertical mural by artist John Gurche, a summary of the chronology of life's journey from the end of the Archean and a quick lesson in geologic time. So they can understand each other, people who study the history of Planet Ocean divide time into three eras: the Paleozoic, or old life; the Mesozoic, or middle life; and the Cenozoic, or recent life. Each era is further divided into periods, which can differ in length by a few million years, depending on subtleties of interpretation. The periods are named after rocks and fossils that turn up in the particular strata of the time. Those names are also applied to the animals the fossils once were, a Devonian fish, for example.

There's no test at the end of *Planet Ocean*, but just so you'll know:

In the Paleozoic era are the Cambrian (570–500 million years Before Present [BP]), Ordovician (500–430 million years BP), Silurian (430–390 million years BP), Devonian (390–340 million years BP), Mississippian (340–310 million years BP), Pennsylvanian (310–280 million years BP), and Permian (280–230 million years BP) periods. In the Mesozoic are the Triassic (230–195 million years BP), Jurassic (195–140 million years BP), and Cretaceous (140–65 million years BP) periods. And in the Cenozoic are the Tertiary (65–1.8 million years BP) and Quaternary (1.8 million years BP–present) periods. For the Cenozoic, since we need more names for things we know more about, the periods are further divided into epochs, from old to new: Paleocene, Eocene, Oligocene, Miocene, Pliocene, Pleistocene, and Holocene. The dominant life-forms in the sea, and eventually on land, for each era and

period are dramatically illustrated in parallel bands on the Tower of Time, with a multicellular colony on the bottom and *Homo sapiens sapiens* on the top.

One of the first things students of geology and paleontology have to suffer through in school is memorizing the names of the eras, periods, and even finer slices of time past, an odious chore at best. So they resort to mnemonics. A mnemonic is a kind of word game to jog the memory on test day and forever more. The better the mnemonic, the easier it is to remember. Everybody has favorites, and here are ours for the periods of the Paleozoic and Mesozoic eras:

CRYING OVER SLEEPING DRAGONS MAY PUZZLE PEOPLE, TERRIFY (OR) JOYFULLY CONVERT

And for the epochs of the Cenozoic era:

PALEONTOLOGISTS EAT ONLY MURKY PLANKTON PORRIDGE HOT

For the details, I sank back into the river of children. They were bunched up in the Ordovician, where one of their teachers was trying to work in a quick lesson on trilobite growth. She was being quite roundly ignored as the mechanics of molting couldn't compete with the television trilobite eye on the wall. The kids elbowed each other out of the way for a chance to stand in front of the camera to see his or her image in the kaleidoscope of screens that represented those astounding, first-ever eyes of a trilobite. Here in the Ordovician, the diversity of the Cambrian narrowed a bit, but trilobites were still king as the period began. Real sophistication waited in the wings, ready to come on stage. By the end of the Ordovician, 435 million years BP, primitive

sponges, corals, starfish, and mollusks had joined the trilobites and earlier life-forms in the shallow seas where light fed the oxygen miracle and warmth encouraged growth. The deep ocean was sparsely populated by drifting algae and other microscopic organisms ready for life's next pulse.

As the hands on the watch of geologic time ticked into the Silurian, everything still lived in the sea, but not for long. (The Silures were a Welsh tribe that lived in the vicinity where the first rocks from the Silurian were found.) The top-dog predators were invertebrates like eurypterids, giant relatives of modern scorpions, some of which were fifteen feet long. Active predation—chasing, killing, and eating something instead of waiting for it to drift into your mouth or gut—really started to catch on in the Silurian. Animals like sponges and corals were doing okay, too, using the old blue-green algae trick of forming reefs. Echinoderms, the five-armed tribe that includes starfish and mollusks, were starting to thrive, among them the first cephalopods, which would eventually become the mighty ammonites. Brachiopods were showing the way for filter-feeders and had refined the body plan they figured out in miniature during the Cambrian almost a hundred million years before. They could now open and close their protective shells. Previously, they were always open, a distinct handicap if something tries to eat you. Near the end of the period (sound the trumpets), fish made their entrance as jawless but mobile vertebrates, the successful children, quite possibly, of *Pikaia*.

Most spectacularly, in the Silurian, living beings came ashore on Planet Ocean after more

IT COULDA' BEEN

LIFE WITHOUT JAWS

OPPOSITE: Eurypterids

than 4 billion years in the water. The first animals left the sea either to find a meal or to avoid becoming one, nobody knows. But the moment came sometime around 400 million years BP, and at the Smithsonian, another animated cartoon and a beautiful diorama draw crowds to witness the event. The landing at Plymouth Rock pales by comparison. According to the Smithsonian's story, the first terrestrial plant was probably a fragile seaweed called *Cooksonia*, possibly a victim of thousands or millions of years of a wet-and-dry cycle in a region of salt lagoons or swamps. To respond to this series of threats to its existence, *Cooksonia* figured out a way to store water in itself with a cuticle, or exterior membrane. Then it developed a form of cellular circulation to move that water, and so became land legal.

The eurypterid terranauts were unlikely celebrities, arthropod nightmares that still form in some memorial section of our brains when we try to imagine monsters. Some were six feet long, with terrifying pincers. In their fossil remains, they look like they were quicker than mosquitoes with bad attitudes. Maybe they were after the *Cooksonia*, or maybe the newly arriving fish were after the eurypterids, who got out of the sea while they still could. Whatever the reason, the land was alive.

When the Devonian began 390 million years BP, the sea boiled with life as the shock waves of the Cambrian explosion rolled on in time. Foremost among its inhabitants were complex invertebrates who enjoyed the top of the heap, while the eventual dominance of *Pikaia*'s children, the backboned ones, was taking shape. The great tribe of ammonites, for instance, was 30 million years into its 330-million-year run on Planet Ocean. The history of life is great theater, with survival getting top billing. No more luminous star than the ammonite ever hit the stage.

EURYPTERIDS

PTERYGOTUS · STYLONURUS

SCOURGE OF OUR ANCESTORS

One of the major roles a species can perform is to survive a mass extinction when a global climate shift, asteroid strike, or other death-dealing event wipes out entire tribes of animals. Hundreds of species of ammonites were masters at this, emerging in the mid-Paleozoic era about 400 million years ago and existing as distinct species for over 330 million years, through several mass extinctions. They finally disappeared at the end of the Cretaceous period, 65 million years ago, when they and the glamorous dinosaurs and about half of all life on the planet also vanished. Before that extinction, newly evolved, shell-crushing marine reptiles like mosasaurs probably reduced the populations and diversity of ammonites to a vulnerable level, no doubt contributing to their demise.

Ammonites were a member of the phylum Mollusca—mollusks—of the class Cephalopoda ("feet-on-head") whose relatives today include the chambered nautilus, squid, and octopus. The evolution of cephalopods seems to have meant getting rid of external shells, though the ammonite shells

and their pumping and buoyancy systems were remarkable. First appearing in the animal kingdom as protective armor, the cephalopod shell eventually proved a unique and powerful adaptation. As an ammonite grew, it left voids or chambers in its shell that were connected by valves called siphuncles, through which it could slowly pass water and air, giving it control over its buoyancy and therefore depth. Evolving an ability to control motion and to rise from the bottom made the nautiloids and then the ammonites the first great predators of Planet Ocean. They were also equipped with beaks to crush the hard-shelled denizens of the ocean floor that were stuck on the bottom with no escape, except to burrow. Trilobites were a main course for the cruising cephalapods, and the fossil record shows their eyes evolved from looking down for forage on the bottom to looking up to defend against an attack from above.

By the Cretaceous, some ammonites were enormous and appeared in either spiral or coiled form. At least one was ten feet long, and shell

many species of ammonites managed to survive the greatest mass dying of all time, the Permian extinction, a much more devastating event than the one that took the dinosaurs (and, coincidentally, the last of the ammonites) at the end of the Cretaceous 65 million years BP. In the Permian extinction at 245 million years BP, for reasons that are not at all clear, an astounding 96 percent of all life bit the dust over a period of about a million years, a snap of the fingers in geologic time. The ammonite's story of survival is so inspiring I bought my daughter a polished specimen at the Smithsonian gift shop and mailed it, with a note, to her:

fragments found in Sweden indicate that some might have reached thirty feet. Predation was in full swing in the sea after 4 billion years of placid algae, bacteria, and simple invertebrates.

Ammonites are very common fossils, like trilobites, because of their hard shells that turn to rock quite readily under pressure. The ability to metabolize calcium and make a shell, and eventually a backbone, was a tour de force of life, a hard trick beside which even the most grandiose evolutionary outbursts like consciousness pale. Ammonites were an improvement over most of their cousins, the nautiloids, because they evolved stronger walls between chambers. This allowed their outer shells to be thinner and therefore easier and quicker to produce. Slow growth can be a distinct handicap in the evolutionary sweepstakes, though not always. The *Nautilus* may even have outlasted the ammonite precisely because it grew more slowly. So much for certainty and design. Somebody just gets lucky or not.

Ammonites were tough, though. About 150 million years into their run on Planet Ocean,

> Dear Laara,
> This ammonite is to honor surviving your first semester at college and to share some of the natural elegance I'm discovering lately. The fossil was found in Morocco, in North Africa, which was once the bottom of the sea. A moment of meditation with it in your hands will give you clues to survival, no matter what the situation.
> Love,
> Dad

IT'S A LONG WAY

*"It's a long way from
Amphioxus,
It's a long way to us.
It's a long way from
Amphioxus,
To the meanest human cuss.
Good-bye fins and gill slits,
Hello skin and hair,
It's a long, long way from
Amphioxus,
But we got here from there."*

—An anonymous paleontologist

(To the tune of "It's a Long Way to Tipperary")

YES, WE ARE ALL FISH

FISH WERE THE FIRST true vertebrates, transforming *Pikaia*'s tentative chordate experiment into the bone and cartilage they would eventually pass on to amphibians, reptiles, birds, marsupials, and mammals. More than half of all vertebrates, extant or extinct, are fish, including five distinct classes: Agnatha, the jawless fish; Placodermi, the now-extinct armored fish; Acanthodii, the now-extinct spiny fish; Chondrichthyes, the sharks and ratfish; and Osteichthyes, the bony fish. Members of all five were alive in the Devonian, known as the Age of Fish during Walcott's time when everything seemed to fit neatly into more comfortable, logical patterns. Invertebrates, too, were roaring along in the finally fertile sea.

The fish evolved like crazy for 140 million years and up into the Pennsylvanian, after which the sharks, ratfish, and bony fish continued the voyage into the future, thus increasing the varieties among them. The senior member of the chordate firm remained, as it does today, the almost-a-vertebrate *Amphioxus*, who goes by the common name lancelet. Sometime before Agnatha, the jawless fish, took the big step and became a true vertebrate, the lancelet, lamprey, and hagfish veered off into a niche that has kept the direct descendants of this primitive life-form alive for almost 450 million years.

With the genetic smorgasbord in full flower, a lot of trial and error went on among the vertebrate tribes up through the Devonian, Mississippian, Pennsylvanian, and Permian periods. The modern ocean, particularly in the abyssal depths, is still full of weird fish. But the creatures of the Paleozoic can really leave you talking to yourself. One of the showstoppers at the Smithsonian, for instance, is a huge glass case around which a steady clump of children stood, stunned by the fossilized armor from the head and shoulders of *Dunkleosteus*, or *Dinichthys terrelli*, a giant of the Devonian that reached the size of the school buses that brought the kids to the museum. Heavy armor plating among fish eventually lost out to speed and teeth, but in its enormous, beaklike, crushing jaws, *Dunkleosteus* could have turned a bus into scrap in a few bites.

By the Mississippian, the sharks were making some real fashion statements with body design, though they left a lot to the imagination. Shark fossils rarely include more than teeth and vertebrae, simply because the rest of the shark skeletons were, and are, made of cartilage. During the Devonian, the characteristics that spell "shark" split their gang into two distinct evolutionary lines that continue to the present.

The Holocephali side of the family includes the noble ratfish and the Elasmobranchii side, the true sharks, skates, and rays. As they transformed themselves over the eons into the sleek swimmers we know today, the sharks took a few wild turns, most notably an amazing creature known as *Helicoprion*, a whorl-tooth shark.

The fossil is common enough that the first one Ray and I ran into was under the bottom shelf in a dusty basement in the Natural History Museum of Los Angeles County, but nobody can figure out quite how it looked in life. Fossil fish man J.D. Stewart showed it to us when we were poking around for Cretaceous fish and reptile fossils. He chuckled as he wrestled it like a hundred-pound stump from its place. "What do you think this guy looked like?" he said. "You never see it in exhibits because nobody can figure out quite how to put it together." *Helicoprion* left its teeth in a perfect spiral, or whorl, on roughly the same plane, with the smaller, older teeth toward the center of the whorl and the bigger, newer teeth to the outside. Maybe it looked like a table saw with fins, or one of those party favors you blow to uncoil a paper tube. Or maybe it had all those teeth rolled around inside an enormous head. Go figure.

While *Helicoprion* and the rest of the primitive sharks were laying their traps for unsuspecting twentieth-century fossil hounds, the Osteichthyes were busy becoming our ancestors. Already, they had split their evolutionary chores two ways. The Actinopterygii, or ray-finned fish, named for their thin, very flexible fins—just like steelhead, a modern example of the ray-fins—never stopped exploding once they got going about 420 million years BP. The ray-fins themselves subdivided into thirds right around the time the Paleozoic ended and the Mesozoic began. They formed the Chondrostei, which now includes sturgeons and paddlefish; the Teleostei, like steelhead and the Cretaceous fossil giant, *Xiphactinus*; and the Holostei, now represented most distinctly by the glorious gar.

In the late Devonian, vertebrates invaded the land on Planet Ocean, but it wasn't the ray-fins that made the trip. The other main stream of the bony fishes, Sarcopterygii or lobe-fins, are the fish that learned to walk, our ancestors. The legendary *Coelacanth latimeria*, the electric blue blast from the past, was named after Marjorie Courtenay-Latimer, who spotted it on the deck of a fishing boat in Africa in 1938. The species was familiar as a fossil, but never before had anyone found a live one, so she passed it on to J. L. B. Smith, a British ichthyologist who described it and became a legend himself. For decades after Ms. Latimer's lucky day, the

coelacanth was everybody's best bet as our immediate finny ancestor, and a four-foot specimen of the great fish is memorialized suspiciously near the mammals in the Smithsonian's story of life.

Lately, the story has changed and most paleontologists are placing their bets on another lobe-fin, *Eusthenopteron*, a rhipidistian. One or the other did the honors, though, and probably in the same way and for the same reasons as the eurypterids, survived wet and dry cycles until it learned to breathe and make its way across hard ground as well as through the water for a meal. Both were well equipped for the adventure. Their fins were muscular, with the beginnings of segments like fingers and toes, and they had developed air bladders, probably for buoyancy, that became lungs.

The Devonian ocean was a tough neighborhood, kind of the South Boston or Asbury Park of the Paleozoic. At the time of vertebrate landfall, slashing teeth, beaks, tails, and fins made survival more uncertain than it has ever been, though the cast aboard Planet Ocean was enormously diverse. Every swimming critter—which was just about everything alive at the time—was part of a gigantic eating and dying machine.

Some inspired pack of rhipidistians came ashore for the first time in the late Devonian, either looking for a meal or trying to avoid becoming one. Maybe they were pursuing the as yet unknowable vertebrate dream of a future filled with cheeseburgers, pizza, Caesar salads, decent wine, homes of their own, and a new car every two years. Most likely, the glorious event was a matter of a single rhipidistian lunging after some unlucky eurypterid out grazing in the *Cooksonia* who ventured too close to the beach at just the right—or wrong—moment.

A TRAGIC END TO AN OTHERWISE PLEASANT DEVONIAN FISHING HOLIDAY.

PALEOZOIC SHARKS

1. **BANDRINGA** *Pennsylvanian*
2. **HETEROPETALUS** *Mississippian*
3. **EDESTUS GIGANTEUS** *Pennsylvanian*
4. **INIOPTERYX RUSHLAUI** *Pennsylvanian*
5. **PROMEXYELE PEYERI** *Pennsylvanian*
6. **ROMERODUS** *Pennsylvanian*
7. **ORNITHOPRION** *Pennsylvanian*
8. **HELICOPRION** *Permian*
9. **POLYSENTOR GORBAIRDI** *Pennsylvanian*
10. **STETHACANTHUS** *Devonian–Pennsylvanian*
11. **INIOPERA RICHARDSONI** *Pennsylvanian*
12. **COBELODUS ACULEATUS** *Pennsylvanian*
13. **ORODUS** *Pennsylvanian*
14. **SARCOPRION** *Permian*
15. **HARPAGOFUTUTOR VOLSELLORHINUS** *Mississippian*

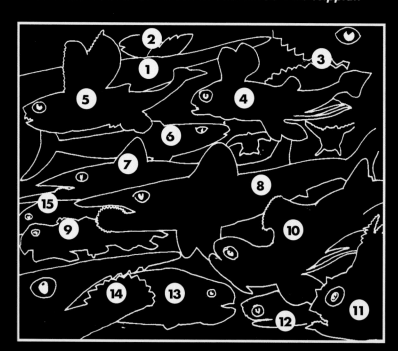

PALEOZOIC SHARKS

PALEOZOIC MIND BOGGLERS

Brad and I were rummaging about in the deepest, darkest corners of the L.A. County Museum of Natural History with paleontologist J. D. Stewart. It was like a scene from some grade-B horror flick, with shelf upon shelf jammed with odd, nasty-looking bones. After hours of prying open crates in this dank treasure house, peering at plesiosaurs and odd cretaceous fish, we were exhausted. As we headed for the door, J. D. slid one last rock out from under a shelf. "Check this out," he said. "It's blown paleontologists' minds for years."

THERE ON THE FLOOR OF THE MUSEUM BASEMENT WAS THE MYSTERIOUS TOOTH SPIRAL.

At first glance, it appeared to be an ammonite. On closer examination, it turned out to be teeth spiraling inward! This toothy vortex had stumped many a scientist. One poor Russian, A. P. Karpinski, spent years in futile attempts to restore the position of the whorl. He placed it in the tail, on the dorsal fin, and in the upper jaw (my favorite, a sinister swimming Dumbo!) of Helicoprion, an enigmatic sharklike fish that flourished in the Permian oceans.

I, too, became obsessed. Eventually my research led me to Dr. Rainer Zangerl, the authority on Paleozoic sharks. Dr. Zangerl assured me that Helicoprion's tooth whorl was "absolutely rigid." I was disappointed because my fantasies of Helicoprion involved a kind of snap-towel, party-favor toothy tentacle.

Living sharks have many rows of teeth and constantly shed worn ones. Helicoprion had a single row of sharp teeth in its lower jaw and only a few flat crushing teeth in its upper jaw. All of its worn teeth were retained in a bizarre whorl, with new large teeth erupting from the back of the mouth.

Edestus giganteus was another shark that Dr. Zangerl described as a "terrible, fierce critter." Its gigantic set of scissorlike jaws sported two single rows of teeth arranged like a shear. I was enthralled.

With Dr. Zangerl's advice, I reconstructed these peculiar sharks (pp. 46 to 47). Few of these fantastic forms survived the Permian extinction.

—R.T.
H. P., P. H., M. F. A.

"MISTER ED"

A FIERCE CRITTER!!

EDESTUS GIGANTEUS

Imagine witnessing the moment from a Paleozoic hunting blind on a dune above a vast mudflat, watching the fish watch the soon-to-be immortal scorpion under a Devonian full moon. Every so often, the water boils and darkens with blood as some nightmare of an animal slashes through the lobe-fins hovering in the shallows, buoyed by their air sacs and taking gulps of the warm, equatorial night as they wait for some edible swimmer to make a mistake and drop its guard nearby. The water settles after one of the attacks, and a bunch of the lobe-fins find themselves driven right to water's edge, and there, scuttling in the beach grass is the eurypterid, an easy target. The lobe-fins may not have hunted ashore until this moment, but they had lived through many wet and dry seasons when their fins grew strong enough to support the weight of their flesh and bones. They self-selected for this trait, since those whose limbs were the most muscular survived, and those that were weak did not. And on this particular, historic night under the killing moon, the fish walked on land and didn't go back. They remained to eat more and probably returned to the sea to breed. But the next generation of young rhipidistians was better off ashore than their parents. The eurypterids and the other invertebrate colonists on land were easy pickings, and compared to dodging armored giants in the ocean, life ashore was a breeze.

Vertebrates are now 350 million years into their run on Planet Ocean, every bit the equal of the other great survivors like Trilobita and Mollusca. Some of each of the backbone tribes have even given up on life ashore and returned to the sea. And some have become doctors, lawyers, paleontologists, teachers, artists, writers, farmers, and physicists. Einstein was a fish. So were Mozart, Virginia Woolf, and Thomas Jefferson. Bruce Springsteen is a fish, and so are Ray Troll, Bill Clinton, and Camille Paglia.

All those children at the Smithsonian are fish, and they might even sense that in the deepest chambers of their subconscious, since they were embryotic mammals much more recently than we old ones. In amniotic fluid identical in salinity to the ancient ocean, they sported gill slits, tails, and, early on, that long, flexible bundle of nerve tissue that makes them chordates, kin to amphioxus, *Pikaia*'s kids. It wasn't hard to imagine them as fish, darting and lunging at each other as they flowed through the galleries, nibbling at the exhibits, breaking water in their most exuberant moments. I could see them pile up as though behind midstream rocks to look at *Dinichthys*, the trilobite eye, or the giant ammonite and hover in the shallows of the dark, silent Archean. Enough of that vision banished the squealing and crying and laughter, but not for long. I still had to yell to be heard on the pay phone in the rotunda when I called Ray to tell him about *Pikaia* after I finally found one of Walcott's breathtaking fossils of our oldest ancestor. It was in a special case, ceremonially alone near the startling explanation of the Permian extinction, the part of the story of life nobody wants to hear.

FISH EATING A CHEESEBURGER

Maybe the rhipidistians were pursuing the as yet unknowable vertebrate dream of a future filled with cheeseburgers, pizza, Caesar salads, and decent wine.

ROLLING SNAKE EYES

DEATH IS AS NATURAL as life, but telling the children is tough. Plants and animals depart Planet

DEATH FROM SPACE AND OTHER BEDTIME STORIES

Ocean in a steady stream as individuals or entire species, even, from time to time, in great mass dyings that have come close to extinguishing all life. Statistically, the leading role of death in the story of life is clearly revealed: At the moment, something like 50 million species are hurtling through space aboard our damp globe. Since the Archean ended in the Cambrian explosion, though, about 50 billion species have put in appearances. The math is easy. But we still have a hard time dealing with the uncertainties that are supplied in enormous doses as the accurate story of life reveals itself in the fossil record.

After the dying ends, the living begins. In the biodebris of a mass extinction, the survivors are left with a clean palette, sometimes free of predators and always shaped by the urgent call to diversify for survival. Extinctions very possibly explain the rough edges in evolution's puzzle and the absence of so-called transitional forms in the fossil record because, during hard times, all life is encouraged to succeed quickly. Though lively debate continues on causes and timing, extinctions are real, and regular people are finally figuring out what they mean in the story of life: Death is the mother of beauty.

Albert Einstein is cherished for having said, among other things, "God does not play dice with the universe." He made that remark as part of his objection to the role of uncertainty in quantum mechanics, which had charmed a new generation of physicists. Many years later, finally, a scientist with a reputation as substantial as Einstein's summoned the gumption to challenge that hopeful sentiment of classical physics. "God not only plays dice with the universe," said Stephen W. Hawking, holder of Newton's Chair in Mathematics at Cambridge, "but

"Exploitation of niches left by extinctions is perhaps the most profound theme in the entire history of life."

—NILES ELDREDGE
Paleontologist, *Life Pulse*

OPPOSITE: Planet Ocean (detail)

**LAND AND SEA DURING
THE PERMIAN**
245 to 290 million years BP

**The greatest of all mass
extinctions at the end
of the Permian claimed
95 percent of all life in the
sea. The Big One also
killed about half the plants
and animals that had
colonized the supercontinent
we now call Pangaea.
Vertebrates, by now full-
fledged mammal-like
reptiles, got a break
and survived to enter
the Mesozoic.**

sometimes throws them where we can't see them."

In the case of mass extinctions, the dice landed under the couch, and uncertainty, luck, and mystery warp our every thought about them. Without a doubt, Planet Ocean has periodically cleaned up its biospheric debris and shows no sign of changing the pattern. For better or worse, extinctions might get rid of mistakes or structural flaws in the evolution of the families of life, but they definitely clear out ecological space for a subsequent outburst of diversity and growth among the surviving species. Some families and species continue through that barrier; some do not. Getting a ticket for the next ride depends as much on good luck as good genes.

The fossil record clearly tells of many extinction events since the Cambrian, of which no single one wiped out all life, though some came close. The celebrated Cretaceous extinction, which took out the dinosaurs, the great marine reptiles, and the stalwart ammonites, claimed about half of all living things. (Paleontologists call it the K-T extinction, for Cretaceous-Tertiary, with the letter K representing Cretaceous.) At the Smithsonian, the literal barrier of a Plexiglas™ wall dramatizes the effects of the extinction that ended the Permian, dividing the Paleozoic, the first great age of life, from the Mesozoic. (Perm is an area in southern Russia after which the period was named.) The bottom two-thirds of the wall graphically depicts a festival of life, a narrow band of species flowing up from the Devonian, Mississippian, and Pennsylvanian and gradually widening to become a swollen bulb of diversity.

By the Permian, life ashore had bloomed to include not only renegade fish and amphibians but honest terran reptiles who took a drink once in a while, breathed air all the time, and bore their young on land. A group of these reptiles, the therapsids,

are even known as mammal-like, and their discovery in widely separated modern sites sparked lively debate on just where the land was during the Permian. Gondwana, the southern supercontinent out of which formed modern Africa, South America, Antarctica, India, and Australia, was heading north in fragments that would eventually join with the rest of the landmass, Laurasia, to form another big continent, Pangaea. Plants had proliferated wildly since the timid arrival of *Cooksonia*. The angiosperms, with their marvelous flowers and seeds, would not bloom for another 170 million years, but runaway forests of horsetails, ferns, seed ferns, and club moss added their oxygen to the venture. These were the coal swamps of the Mississippian and Pennsylvanian, the remains of which we now use to run trains, planes, cars, and furnaces.

Behind us, as we face the Smithsonian's extinction wall, is the counterpoint to this greatest of all extinctions, a Permian reef. Meticulously constructed by museum preparators from tens of thousands of minuscule components, the reefs clearly cry, "Life." When they occupied vast tracts of the shallow seas 235 million years ago, their dimensions would have been measured in hundreds of miles. The Permian reefs were epic constructs of sediment formed from the remains of crinoids, the ubiquitous, calciferous plantlike animals also known as sea lilies, other echinoderms, brachiopods, rugose corals, mollusks, arthropods, and formiferans, all stabilized and bound together by bryozoans, chambered sponges, and red and green algae. The reef is orange, green, and yellow in its dominant hues and speaks fluently of balance and complexity.

Around these enormously diverse communes, fish and other free-swimming members of the cast

that both nourished and fed on the reef flourished, hung in the diorama by invisible threads. Predators and prey engaged in their tragedies of balance, odd worms and arthropods staked out holes in the faces of the reef, and the seafloor itself was a quilt of life. The familiar trilobites still scuffle around like hungry bumper cars, though in smaller numbers than before the last extinction at the end of the Devonian claimed so many of them. The siphons of the filter-feeding brachiopods and other burrowers make their tentative incursions from the mud, and the crinoids, looking more like vegetation than the true animals they are, make dense subsea forests at the peak of their dominance. In time to the rhythm of the water washing against the reef, the tendrils of oxygen-making seaweeds perform their photosynthetic hit song and sway like the bows in the violin section of an orchestra. Modern reefs, though magnificent, are weak successors to the Permian temples of evolution.

But something happened.

For reasons still not completely clear in the tantalizing story of Planet Ocean, more than 95 percent of all life disappeared within a mere million years or so. The top third of the Plexiglas™ wall of extinction at the Smithsonian is no longer a dense cluster of plants and animals but an almost clear window on the future. Through it, you can see into the next gallery and the Mesozoic, where the wave of life begins to rise again from the few survivors. Trilobites and very nearly every marine invertebrate disappeared, as did members of every phylum. A few species of ammonites and brachiopods made it. So did some fish, reptiles, and amphibians, keeping vertebrates in the picture. As the paleontologist Steven Stanley says, "Things got bad. Then they got worse."

For a couple of centuries, paleontologists have known that when rocks with a lot of fossils lie under or over rocks with very few fossils, it means something big happened to kill off a lot of life. The Paleozoic, Mesozoic, and Cenozoic boundaries were drawn in the mid-nineteenth century by the geologist John Phillips, in fact, because the rocks told him that life ended and began anew at least three times on Planet Ocean. His contemporary, Charles Darwin, incorporated extinctions into his theory of steady improvement through competition by arguing that the mass dyings were gradual

affairs. He, of course, claimed that the rocks supported him, too.

The fossil record contains gaps; there are few transitional animals to tell us what happened between, say, a soft-bodied invertebrate and a trilobite. The absence of transitional animals has been among the chief arguments of creationists, who claim a series of God-driven events, not evolution, explains life-as-we-know-it. Stephen Jay Gould and Niles Eldredge, though, have successfully revised, rather than dismissed, Darwinian logic by explaining how speciation can occur in bursts, with relatively stable, long stretches of time between such events. According to their theory, called "punctuated equilibria," the gaps in the fossil record are real, but the absence of transitional forms is the result of the relative brevity of the periods in which they are actually in the state of change, therefore leaving fewer traces. The tools for dismantling the inconsistencies of gradual evolution in favor of a new and better story are found in extinctions.

But how do they happen? A good mystery is a feast for generations of scientists, and every fossil hunter, cosmologist, and biologist seems to have taken a shot at explaining extinctions. Many more than three mass dyings have altered the flow of the stream of life, and most extinction theorists are selling two likely, somewhat obvious, causes: The planet itself did the killing with a relatively sudden change—say, over a million years—in the climate, atmosphere, or another ingredient that balances the ecosystem; or something extraterrestrial, like an asteroid or a comet, slammed into Planet Ocean, and a few very bad weeks of aftermath did the job. A combination of the two is the front-runner, but extinctions have been glamorous since the eighties because of the charm and theatrics of the death-from-space theories.

Imagine you're driving home from work one day, listening to National Public Radio and letting your mind drift through the weary logistics of life in the nineties, wondering whether the disenchantment you feel for your job will disappear, or if maybe you should look for something else to do. You glance over at the driver next to you, who's matching your sixty miles an hour. Mere inches separate you from a hideous collision if either of you, for even a moment, lapses from the rhythm of fast, heavy traffic. Then your mind settles back on the radio and the nasal drone of a bulletin from the latest war. Finally, the science reporter tells you that an asteroid will pass within the moon's orbit with long, but nonetheless very real, odds that it will slam into Planet Ocean. You're used to this guy's gee-whiz stories of discovery, but his voice has taken on the stridency reserved for hotel fires and plane crashes. He actually sounds like he's on the verge of panic.

And why didn't a renegade asteroid make the top of the news? Because nobody can deal with the real possibility of a big earthquake, never mind a hunk of cosmic debris ten miles in diameter piling into Nebraska, Paris, or the Caribbean. The effects of such a dark moment would extend far beyond the impact. In all likelihood, Planet Ocean would be set afire, water and even rock would vaporize, and a long, long night would begin. The absence of light would probably do the job on plants and animals not in the immediate vicinity of the blast, a much more hideous version of the now-familiar apparition of nuclear winter that made Ban-the-Bomb fans of even the most bellicose.

Mass extinction from space, once a backwater of evolution esoterica, was thrust into pop culture in the last quarter of this century by Luis Alvarez, a physicist, and his son, Walter, a geologist. Together,

THERAPSIDS

These reptiles start showing up in the fossil record during the Carboniferous, 275 million years BP, when they broke from their amphibious predecessors. Therapsids, our reptilian ancestors, share a distinct skull design with mammals. By the Triassic, 240 million years BP, therapsids may even have sported body hair, and soon after evolved into true mammals.

THE FAMILY TREE

AN ARTISTS VERSION OF

WITH A RATFISH BIAS *

Within a mere million years or so, more than 95 percent of all life disappeared in the Permian extinction.

they figured out that a thin layer of rocks near Gubbio, Italy, contained more than thirty times the normal amount of a certain rare element called iridium. Not incidentally, that layer of rock was also the K-T boundary, the exact rock that was the mud, soil, and organic debris at the end of the Cretaceous. In Gubbio and other places, the Alvarezes knew, a rich fossil flora and fauna including dinosaurs are found in the rocks on the Cretaceous side of the boundary and very few fossils of anything at all on the Tertiary side. More critical to the theory of relatively instant annihilation, the fossils of plants, pollen, and spores clearly reveal sudden devastation and, after the event, recovery.

The K-T boundary itself was old news. But the new interpretation of the iridium layer was a bombshell. The telltale element is only found in any quantity at all in the core of Planet Ocean, but it is common and concentrated in meteors and asteroids. The iridium could have come from volcanoes, and many geologists believed in the volcanic theory over death-from-space for a long time before the Gubbio breakthrough. But Walter Alvarez demonstrated convincingly that the Italian rock was not volcanic at a time when the iridium layer could have been deposited. And within a couple of years, when other investigators started taking death-from-space seriously, the K-T iridium layer showed up all over in just the right places, along with enormous craters and open minds.

We know that asteroids, comets, and big meteors have participated in mass extinctions of life on Planet Ocean. Some paleontologists are convinced that an asteroid, perhaps more than one, played a lead role in the Real Big One, the Permian extinction. Some of the mass dyings have been bigger than others, but it seems like they are on a

26-million-year cycle, whether by chance or some astronomical rhythm about which we know nothing at all. Some have proposed that our orbit around the sun takes us periodically through cluttered sectors of space, producing a collision or two, or that the orbit of a comet or other celestial body has intersected with ours in an even more complicated system. If 26 million is a good number, we're due in 12 million years. But there's no reason to take comfort from these statistics. Smaller mass dyings that make the black death seem tame happen much more often than once every 26 million years. And most extinctions are best explained by a combination of gradual species decline with a spectacular coup de grâce from the cosmos.

For some reason, death-from-space is a more popular theory among ordinary folk than that of the "normal" and gradual decline and disappearance of individuals, species, and entire phyla from the record of life. Maybe it's the show-biz appeal of an asteroid strike, the familiarity of act-of-god scenarios in our myths, or the complexity of ecosystems that hide cause-and-effect from us simpleminded vertebrates. Planet Ocean itself does a lot of killing with its shifting climate, magnetic poles, and migrating water and rocks. A million years of decline is not an instantaneous event. Neither is it a substantial hunk of the entire history of life, nor even the history of life between mass extinctions.

Without the iridium layer and some newly discovered evidence of impact craters, evolution through the flow of extinctions would remain the prize of the gradualists. We know that the positions of the landmasses shift constantly, creating sometimes stable and at other times disastrous weather patterns. We know that the ice comes and goes,

altering sea level and changing the components of the atmosphere. We know that by the time the asteroid or comet ended the Cretaceous, only a few species of the great marine reptiles and dinosaurs were still around for the fireworks. Most had died off in a steady decline from their heyday in the Jurassic, between 120 and 70 million years before. When the trilobites caught the Big Bus in the Permian extinction, they had been hanging on by just a thread of species for almost 100 million years. They were sitting ducks for an asteroid or just a bad few thousand years.

The mass dyings in current catastrophist theory are not the relatively minor floods and fires that gave the same name to the nineteenth-century catastrophists. The term is now applied to those who think extinctions are among the dominant themes in the evolutionary song. Gradualists are their opposites, those set in the scratchy groove of steady progress, whose ranks are thinning as the evidence mounts. Darwin probably would have bristled at the cosmic punctuation marks of modern extinction theory and raged at the wild luck involved in making it through one of those barren epochs, but he would most likely have come around in the end. Still, it's hard to fit the general truths of mass extinctions into freeway driving, steady jobs, and the approaching moment of our own departures from Planet Ocean. Survival is at least as much a matter of chance as fitness, so you can eat right, get plenty of exercise, run out grounders, pledge allegiance, recycle your trash, keep your eyes on the road, and still get whacked for no reason at all.

Nature is not fair.

WHO'S WHO IN LIFE'S FAMILY TREE

KEEPING TRACK OF 50 billion different kinds of plants and animals, dead or alive, is the kind of chore that drives a lot of folks away from science. People who classify life-forms are called taxonomists. Their trade, taxonomy, summons images of meticulous, plodding curmudgeons embroiled in the controversies of crediting discovery, honoring the discoverers with their names or the names they choose, and declaring who's related to whom. The naming is only part of the job, because you also want a system of biological naming to reflect past relationships in the evolutionary flow and to say something about likely futures. A lot of dust has been kicked up deciding just what sort of classification best describes the one true tree of life.

You can classify living things according to endless kinds of systems. Ray Troll, for example, can be classified as a middle-aged male human with limited fishing skills who draws terrific pictures and who lives in the Milky Way Galaxy, near the Sun, on Planet Ocean, in North America, Alaska, and Ketchikan, on a hill. That arbitrary assemblage of characteristics and geographic detail would help you find Ray if you're coming from Alpha Centauri. You could also have his street address and zip code, or the name of a friend who can just point to his house. There is, however, only one correct way to locate Ray according to his ancestral relationships, a system called cladistic taxonomy, which is not arbitrary at all.

The word *clade* is from the Greek *klados*, meaning branch, and the classification of organisms as they have evolved through time is based on the concept of successive branching from common relatives, some extinct, some alive. Eukaryota, Animalia, Deuterostomia, Chordata, Vertebrata, Mammalia, *Homo sapiens*, for instance, is a list of the branching points of some of the relatives we humans have in common, our evolutionary address. Though this list can be further refined to describe the clades we share with other primates, early mammals, and specific kinds of vertebrates like fish, it is perfectly accurate at every level of detail. Taxonomy is not just a catalog, like other systems of classification can be, but rather a history of the evolution of life that began only once, survived many mass extinctions, and changed form billions of times.

What does a Zen monk say when he orders a hot dog? "Make me one with everything." That's an old joke, but cladistic taxonomy makes it a truth. It's difficult for most people to understand that we are absolutely related to chimpanzees, but that we did not evolve *from* chimpanzees. Instead, we have a common ancestor with whom we shared the same branching point something like 5 million years ago, much more recently even than the branching point chimps share with orangutans. More difficult to grasp are the branching points we share with animals that are extinct or hundreds of millions of years older than our tribe, like coelacanths, amphioxus, and other living fossils. All mammals, including humans,

whales, dogs, cats, and horses, are related to fish at exactly the same branching point but not in any hierarchic order after that. Pre-Darwinists incorrectly assumed that humans were at the top of what they called the Great Chain of Being, which placed humans farther from fish than from cats, for instance, and cats somewhere in between. But those weird creationist drawings of so-called intermediates like half-dog, half-cat are wrong mainly because they do not use the correct system for classifying life-forms.

You can argue about relationships and about the precise moments of branching, lately an indoor sport that is tying up a lot of paleontologists with the possibility that birds are living dinosaurs that somehow survived the K-T extinction. Birds and dinosaurs shared a branch and split either before or after they became bird or dinosaur. If they split before, birds are no more dinosaurs than mammals are fish. If they split after, birds, in effect, evolved from dinosaurs and are, therefore, dinosaurs. If you want a lively evening, bring this one up during the cocktail hour at a paleontology conference. But wait until we're in the Mesozoic so you'll know what to ask. One thing for sure, birds and dinosaurs are both related to fish, just like the rest of us vertebrates.

AN EXTINCTION FUGUE

LIFE IN THE LIMESTONE mines of Holland in 1770 was no trip to the beach, though the valuable rock, in fact, was the seafloor during the Upper Cretaceous, 65 to 80 million years before. The miners hadn't a clue, of course, but during their long, hard days of digging and hauling they often found bits of what looked like shells and bones. In a cavern near Maastricht on the Maas, or Meuse, River, a crew of those anonymous miners came

What does a Zen monk say when he orders a hot dog? "Make me one with everything."

Convergence

Chateau Mosasaur

across a set of jaws three feet long. The story of the first mosasaur is probably the most often-told in paleontology, no doubt embellished by each worthy teller, but it goes something like this.

The miners rushed from the dark pit with their discovery and straight to a German military doctor named Hoffmann, who was known as a generous buyer of the odder bones and shells. Hoffmann paid the men, took the jaws, and promptly got sued by one Canon Godin, who owned the land beneath which the mines lay and who wanted the bones for himself. A local judge agreed, and in short order, the jaws passed into the sanctum of Godin's château, where they remained on display in a glass case for twenty-five years. They became famous. Folks at the time were fascinated by relics, most of them of religious significance, and speculated that the giant animal who sported the jaws must have lived prior to the great flood and missed Noah's Ark for some reason. That was as close to extinction theory as anybody got in those days. But the jaws were not the only puzzling bones coming out of the ground. From the Paris plaster mines and subway excavations, other miners were pulling the remains of giant elephants—or so they thought.

In 1795, Napoleon's Republican armies were laying waste to Europe. As it happened, one of them was shelling the town of Maastricht. So famous was Canon Godin's relic from the limestone mines, though, that their general, Pichegru, ordered his cannoneers to spare the château. In the aftermath of the battle, while the traditional looting was under way, it turned out that Godin, fearing for the safety of the jaws, had hidden them elsewhere in the town. The looting and plundering was a grim and nasty reward for the troops, many of whom were paid nothing except for what they could seize after taking a town. Pichegru coveted the jaws as a trophy of war and turned up the volume on the party by offering a reward of 600 bottles of fine wine. In short order, he had them. The giant jaws from Holland returned with the conquerors to Paris, where they forever changed the way we think of the past.

Paleontology began in the City of Light, where naturalist Jean-Baptiste Lamarck hammered out a full-blown theory of evolution at the end of the eighteenth century from the fossil-rich rocks of the Seine valley. His work cleared the way for Darwin's more refined version and set the stage for the first story of life based on fact rather than myth. Lamarck's rival, Baron Georges Cuvier, was the first to examine the yard-long jaws, complete with huge teeth, of the fossil he called the "Lizard of the Maas" thirty years after its discovery in the limestone caverns of Maastricht. He could no longer deny the obvious, and stunned the members of this new scientific discipline and eventually the world with the proof that all is not as it ever was. Nothing like this giant animal, later named mosasaurus, was alive on earth at the time. Cuvier was a powerful champion of the creationist wing of paleontology, with its theories of limited time for making life, so it is ironic that he became the messenger who declared that some animals have become extinct, some remained, and new ones arrive from time to time. Extinction is real.

Extinction became real when Napoleon's army brought the plundered mosasaur jaws back to Paris.

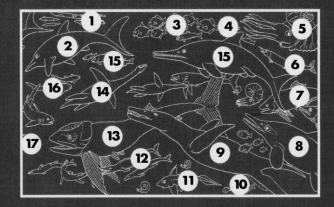

THE JURASSIC SEA

1. **BELEMNITE**
2. **ICHTHYOSAURUS**
3. **MESODON**
4. **PHOLIDOPHORUS**
5. **AMMONITE**
6. **ASPIDORHYNCHUS**
7. **HYPSOCORMUS**
8. **OPHTHALMOSAURUS**
9. **LIOPLEURODON**
10. **DAPEDIUM**
11. **ISCHYODUS**
12. **METRIORHYNCHUS**
13. **LEEDSICHTHYS**
14. **CRYPTOCLEIDUS**
15. **LEPIDOTES**
16. **HYBODUS**
17. **SPATHOBATHIS**

The Jurassic Sea

She Sells Seashells

MESSING AROUND IN THE MESOZOIC

MARY ANNING AND MARY ANN

ONE DAY IN 1810, on a beach beneath the chalk bluffs of England's southern coast, an eleven-year-old girl found an animal skeleton the size of a railroad car that nobody had ever seen before. Mary Anning and her brother, who remains nameless, were collecting fossil seashells, which they sold to support themselves and their widowed mother, when they came across huge bones in the face of a cliff. At her age, Mary was hardly a member of the nascent tribe of paleontologists, but she carefully chipped away the soft chalk to expose the remains of the ancient beast. Real paleontologists and other scholars, known in those days as savants, flocked to the beach at Lyme Regis for a look at this shocking treasure in the shale. They found it to be fully articulated, with its head, needlelike teeth, backbone, ribs, tail, and four paddle-shaped flippers all in their proper places.

The savants eventually named it *Ichthyosaurus*, literally "Fish Lizard," because it looked like a combination of both kinds of familiar modern animals. Within a decade, everybody in Europe had seen a drawing of the creature, and Mary Anning became famous enough to inspire the tongue twister about her seashells.

Alive, ichthyosaurs had introduced high voltage into the evolutionary dynamo of predation. The Mesozoic began with the silence of the early Triassic, gradually yielding to the rumblings of life a few million years after the Permian extinction spared only a few fish, some lucky invertebrates, and plants in the sea. Ashore, the flora and fauna were similarly distressed, and some of the surviving reptiles apparently said the hell with it and went back to the sea

"She sells seashells by the seashore."

—ANONYMOUS

MARY ANN FOUND AN IGUANODON TOOTH.

Mary and the Fish Lizards

MARY AND THE FISH LIZAR

A farmer named Alfred Leeds found the first fragments of this middle Jurassic giant fish near Peterborough, England, in 1889. Paleontologists named the creature *Leedsichthys problematicus,* mainly because the fish was so huge nobody could imagine it alive. The body parts Leeds found, and others that turned up in Germany, would have belonged to a fish over a hundred feet long, longer than a modern blue whale.

where the pickings were a little easier. Ichthyosaurs evolved anatomies and instincts that capitalized on what became an explosion of fish, and they tuned themselves perfectly for the seafood diet. By the end of the Triassic, several species and sizes of ichthyosaurs were swimming in the sea, all of them sporting the dangerous-looking snouts familiar to anybody who has ever seen a gar or pike. Their stiletto teeth were perfect for grasping a slippery prey, and their long jaws streamlined their bodies for the critical talent of swimming fast in short bursts. Ichthyosaurs were dead ringers for modern dolphins, but they weren't mammals and their tails were vertical instead of horizontal. They were the first of the giant marine reptiles, classic examples of

convergence: Form follows function, especially when it comes to getting a meal, and evolution repeats body features to take advantage of the same niches in the food web through eons of time. If fish are all over the place and you have a gar snout and a dolphin body, you're in Fat City.

The refinements of convergence and other revelations of the fossil record were still beyond the grasp of Mary Anning and the English savants of the early nineteenth century, but the dice were rolling. When you read the history of paleontology, it's easy to get the impression that women have all the luck when it comes to earthshaking discoveries, but they end up either with a debatable campfire yarn or with no credit at all. The heroines include

Helena Walcott and the Burgess shale, Mary Anning and her sensational ichthyosaur, Marjorie Courtenay-Latimer and her electric blue coelacanth, and Mary Ann Mantell, who took a walk down an English country lane in the spring of 1822 and found the first dinosaur.

Cuvier's Lizard of the Maas and Mary Anning's ichthyosaur transformed an entire generation of Europeans into fossil hounds. Few, however, were making the evolutionary connections that would later enshrine Charles Darwin. At the time, less than a billion humans were alive on Planet Ocean and their tools for excavation and preparation were primitive. So they encountered nowhere near the number of fossils we do today with our 5.5 billion people, jackhammers, dynamite, dentist drills, and fine microscopes. For the nineteenth-century fossil hunters, the stories from the rocks were best-sellers, front-page news that inspired amateur and professional paleontologists, whose explanations for the bits of teeth and bone they found were wildly incomplete but whose enthusiasm was nothing short of rapturous. Among them was the English surgeon Gideon A. Mantell, the acknowledged father of dinosaur mania.

Marine reptiles such as mosasaurs, ichthyosaurs, and plesiosaurs were not dinosaurs, though they evolved in the sea to enormous sizes at roughly the same time that their distant cousins, the dinosaurs, were thundering around on land. Both had common ancestors at the point where the reptile branch split from the amphibian branch, and Dr. Mantell and his wife, Mary Ann, started us on the way to realizing that big lizards ran the show ashore during the Mesozoic. The Mantells shared their paleohunger and had even published catalogs of common marine fossils, so they probably took advantage of every

chance to poke around in the chalk, shale, and sandstone. Dr. Mantell was making a house call in Sussex, according to the most romantic version of the tale recorded by the doctor himself, and he left Mary Ann in the carriage to bask in the spring sunshine of the Ouse valley north of London. She got restless, took a walk, picked up an enormous tooth from a pile of rocks beside the road, and, when Gideon finished his medical chores, dropped the celebrated fossil in his hand. Mantell's reasons for encouraging this version of the discovery moment are unclear, and the account has been disputed by many, including John Noble Wilford in his fine book, *The Riddle of the Dinosaur*. No matter what really happened, within a year Mantell had sent the tooth and some related bones to Cuvier in Paris in the hands of none other than Charles Lyell, whose theory of geologic time would eventually replace the seven days of biblical Genesis.

By this time, Cuvier was the *grand fromage* among the scientists of Europe, a pompous man resting on the laurels of his earlier work. He sent the tooth back to Mantell with his brusque pronouncement that it was nothing more than part of a rhinoceros. Mantell was daunted, but by several accounts, he remained obsessed by his fossil's inconsistencies with either the known marine reptiles or modern land animals. By 1825, he was sure the creature had been a large, plant-eating land reptile something like an iguana, and he finally published his conclusions, naming it iguanadon, meaning "iguana tooth." Mantell was wrong about a lot of things he surmised from the fragments of his iguanadon, but, like Cuvier, he gave us permission to interpret the messages of life in the fossils in an adventurous new way.

Sixteen years later, another physician, the

LAND AND SEA DURING THE JURASSIC
140 to 210 million years BP

The middle period of the Mesozoic, the Jurassic, popularly known as the Age of the Dinosaurs, also featured oceans of fish, invertebrates, and reptiles. The skies were full of flying reptiles, called pterasaurs, and real birds made their debut. So did lizards, earwigs, and redwood trees. But the place was anything but a park.

THE·DATA·IS

**The Data Is
in the Strata**

N THE STRATA

xiphactinus

Sword Ray —

"Whatever it might cost me in privation, danger, and solitude, I would make it my business to collect facts from the crust of the earth; that thus men might learn more of the introduction and succession of life on our earth."

—CHARLES H. STERNBERG
In a letter to E.D. Cope, 1876

anatomist Richard Owen, would coin the term Dinosauria—Terrible Lizard—to include iguanadon and a handful of other fossil reptiles, which by then were just pouring out of the quarries and cliffs of Europe. Sadly, Mantell was so driven by his work on iguanadon and the rest of the critters of the chalk that he neglected his medical practice and his family. In 1839, Mary Ann took the kids and left him to his rocks and bones.

BADLANDS, GREAT FOSSILS

RAY AND I CAME UPON the badlands of Alberta late in the day from the west, just as the the sun behind us was shocking the wheat fields into lurid postcard yellows. Ray was a mile ahead of me on an otherwise empty two-lane, and though the scenery was symphonic, nothing even remotely resembling the gnarly canyons and hoodoos of late Cretaceous chalk was anywhere in sight. Then, just as I was thinking Ray had blown the navigation and missed a turn, his truck vanished like it had dropped off a cliff in a cartoon. A minute later, I followed him down a Hail Mary grade, skidded into a gravel pull-out and leaned, stunned, on my car door looking down at the Red Deer River and the badlands dancing in the prairie twilight. At last, we were in Dinosaur Country.

For a year, Ray and I had been reading and swapping stories about the rush to the North American fossil beds, and we stood speechless in the evening calm above the Red Deer as the dimensions of reality slipped into the legends of George Dawson, Barnum Brown, Charles Sternberg and his sons, Othniel Marsh, Edward Drinker Cope, and the other pioneer time travelers. Their lives coincided with the few decades when the great

vertebrates of the Mesozoic came alive again in the halls of the great museums to fascinate people all over the world. The Golden Age of Discovery spanned fifty years, beginning in about 1870 after the European paleontologists handed *Homo sapiens sapiens* a new glass through which to view extinction and the remains of the past.

If you want to find fossils, dig where they are, as Casey Stengel might have said. And where they are depends on what was wet or dry and when. If you are looking for marine reptiles from the Triassic, for instance, you have to find sedimentary rock like the English shale that was underwater when those animals were alive. If you want to find land animals, like dinosaurs, you have to find rock that is near the surface today and that was dry or swampy when they were alive. The Mesozoic (Terrify or Joyfully Convert, remember) lasted from 230 million years BP to 65 million years BP and was prime time for big vertebrate fossils. What are now the rocks of North America are a mélange of what were just the right conditions during most of that 165 million years and now revealed in the strata. The fossil beds of the Alberta badlands, for instance, were part of a continent during the late Cretaceous, a lush zone of coastal deltas bordering an inland sea to the east. The rippling prairie overlays even earlier strata, which accounts for the nodding donkey pumps sucking oil from the shales of the Mississippian and Pennsylvanian (May Puzzle) a few hundred feet beneath the surface. Just south, in Montana and Wyoming, most of the land was dry during the Jurassic and the Cretaceous. East of there, during the Cretaceous, a shallow inland sea occupied a basin that is now the middle of the modern continent, so you find the fossils of marine reptiles, flying reptiles, and fish from that period. But no dinosaurs.

The terrain of the Alberta badlands just *looks* fossily, in the same way that parts of a stream can look fishy. Erosion is a paleontologist's pal. The first thing you look for, even if you don't know what geologic period you're in, is some feature that breaks up the surface. A river is a great eroder; so are wind and runoff water on steep slopes, exposed cliff faces, and the occasional big flood, like the ice-age deluge that carved the badlands. Curiously, automobiles have been a boon to fossil hunters since a lot of the geology and paleontology in North America goes on in road cuts. Quarries are good, too, whether they're for mining coal, gold, or gravel, though sometimes they're dug and blasted just to excavate fossils.

The badlands were so named because the earliest arrivals—the nomadic ice-age bands, Plains Indians, and, later, European settlers—figured the dramatic but inhospitable landscape was good for absolutely nothing when it came to gathering or growing food or feeding livestock. What's more, the desolate, twisted ditches, caves, canyons, and queerly eroded pinnacles we call hoodoos are full of snakes and scorpions. The modest but nicely balanced flora and fauna also include prickly pear cactus, bull thistle, western meadowlarks, red-shafted flickers, coyotes, and rabbits. The Sioux called the place *maka sika*, which means land bad, and French traders used *la mauvaise terre*, literally, bad land. The early people would be surprised to know that this scarred patch of the continent has been declared a United Nations World Heritage Site (like the Burgess shale a few hundred miles west), a treasure held in trust for the entire world. Nobody seems to know what the Indians thought of the ghostly bones and teeth that littered the landscape after a hard spring rain.

Charles Hazelius Sternberg, the son of a pioneer Lutheran preacher, knew the badlands and their bones were priceless. In 1912, at the peak of his spectacular career as a fossil hunter, he literally barged into the country from the northwest, choosing passage on the Red Deer River over yet another gut-wrenching journey by horse and wagon. He had been hired by the Geological Survey of Canada after aggressive American collectors had angered the Canadians, who didn't want "their" fossils ending up in the museums of New York, Washington, and Boston. Barnum Brown, for instance, had rushed to Alberta after a tip from a rancher who, on a visit to the American Museum in New York, said something like, "Those bones you guys are arranging into giant skeletons are all over the place out where I come from." The rancher was from the Red Deer River valley.

The spirit of the times was decidedly competitive, masculine, and invigorated by public celebration of great rivalries among inventors, scientists, explorers, and financiers. Discovery was the nineteenth-century equivalent of modern television sports, though of infinitely greater substance. Newspapers thrived on accounts of grueling sprints to the North Pole, the feverish slaughter of the buffalo and the Indians, and the race for the dragons from the ancient past. The old-time fossil hunters wore seasons in the Kansas chalk like badges of honor because there, during the decade that straddled the

Ichthyosaurs were dead ringers for modern dolphins, but they weren't mammals and their tails were vertical instead of horizontal.

From the jumbled bones in the ancient mud, a Parasaurolophus *rose as though in a home movie run backward, reassembled itself in the sky, slipped into its camouflage skin, and settled, alive, to the ground.*

American Civil War, Edward Drinker Cope and Othniel C. Marsh mounted their legendary paleontological campaigns for their museums and reputations. Their antagonism and competitive juices fueled the greatest fossil rush the world has known, as Cope, Marsh, and legions of other less-celebrated collectors swept northwest through Colorado, the famous Como Bluffs of Wyoming, and the great dinosaur beds of Montana.

Meanwhile, at roughly the same time Cope and Marsh were locked in paleocombat in the United States, more modest endeavors were under way in Alberta, where the British Empire delivered another physician/fossil hunter, Dr. George M. Dawson. He was a tiny man, barely five feet tall, who had abandoned medicine and taken a job mapping the boundary between Canada and its southern neighbor on behalf of the Queen of England, Ruler of the Commonwealth, Empress of India. Dawson and his assistant, Joseph Tyrrell, a reformed lawyer, discovered the mother lode of fossils in the Red Deer River valley when they excavated the near-perfect skull of a huge carnivorous dinosaur, now called *Albertosaurus*. If you called central casting in Hollywood and said, "Send over a prehistoric monster," you'd probably get something like *Albertosaurus*, a slightly more compact version of the familiar *Tyrannosaurus rex*, a big, fast, top-of-the-food-web creature with teeth that define fright.

Often, the protégés of great rivals continued the fray or clashed horns and testosterone with other fossil hunters. So it was that in the Alberta rush, Charles Sternberg, once apprenticed (though begrudgingly) to Cope, met head on with Barnum Brown, a former student of Samuel Williston, the Kansas King, who was himself apprenticed to none

other than Marsh. Cope and Marsh were long dead by the time Brown and Sternberg reached the northern coasts of what had been the Western Interior Seaway and went bone hunting in Alberta. The place was so rich with fossils they couldn't drive a tent stake without discovering something. Both soon were shipping magnificent specimens of horned, duck-billed, armored, and carnivorous dinosaurs. From their separate expeditions, against all odds, one of Sternberg's three sons, George, found pieces of mummified dinosaur skin. Then so did Brown, adding enormously to our tentative collective vision of the late Cretaceous. The bones, teeth, and skin told them that herds of great beasts once harmonized in life as predators and prey, somewhat like the animals of modern savannas. They even found some creatures locked for the ages in the final embraces of mortal combat—the jaws of a squad of carnivorous, jackal-like dromaeosaurs, for instance, fixed to the vertebrae of a duck-billed hadrosaur. In places, they found bone piles so thick the only explanation was a big killing of some kind, not a mass extinction but maybe a nearby volcanic explosion. The blast could have stripped the meat off every living animal. Then a subsequent and likely flood washed all the bones into a mass grave downstream. Sixy-five million years later, paleontologists would find this jumble of fossils where thousands of dinosaurs had died, as they did along stretches of the Red Deer River.

With the tenacity of gold prospectors, the old-time fossil hunters hung in there against the elements, desolation, and loneliness. They put in miserable but momentarily thrilling days with pick, shovel, and buckboard. And they raced each other, together with the wind and water they knew would in time destroy what they didn't extract, wrap in

plaster, and haul to a railhead. According to one account, a collecting party came upon a particularly fine dinosaur late in the season after their food had run out. Rather than abandon their fine specimen, they ate their horses and manhandled the wagon over the hard miles to their riverboat.

We made our own camp where Little Sandhill Creek meanders into the Red Deer at Cottonwood Flats, barbecued hamburgers, and watched the moon rise over the badlands. Though the flood of summer tourists to Dinosaur Provincial Park was still a month away, the night songs of the frogs and coyotes mingled with sounds of sputtering lanterns, beery chatter, and TV sets from the RV jungle. Still, visions came easily of Brown and Sternberg, hunkered down with their families and assistants in about the same place, savoring the wilder sounds and aromas of life in the field. They, too, probably tossed and turned through sleepless nights early in their seasons, unable, as I was, to banish the specter of a Cretaceous delta 75 million years ago.

That night in a tent on the banks of the Red Deer River, I couldn't sleep because dinosaurs rose up from the badlands and rattled around in my mind. Sleeplessness is not an unusual condition. I, like you, perhaps, have been kept awake by the sadnesses of a child, the anxieties of love, and the more general fears of tax men and bogeymen. But that night, real dragons roamed outside my tent, and I could see them, smell them, and hear them trumpet, whistle, and roar. From the jumbled bones in the ancient mud, a *Parasaurolophus* rose as though in a home movie run backward, assembled itself in the sky, slipped into its camouflage skin, and settled, alive, to the ground. Then dozens of them materialized, a herd, ambling like enormous Halloween cows toward a slow, brown river for evening water,

with young ones prancing along, yearlings in this early spring. The newest hatchlings were still hidden and guarded in the scrub by adults who had drunk earlier and now defended their eggs like birds, decoying predators if at all possible, avoiding confrontation and revelation of the nest. Chasmosaurs and centrosaurs filtered in, too, from the more defendable open flats to the closer quarters of the river where they were vulnerable to attack from cover. The lumbering giants howled and hissed at each other, defining territory, bobbing and quivering as their eyes, ears, and, perhaps, the sense receptors in their flaring head frills searched for the slightest threat. The *Parasaurolophi* bellowed their slide-trombone notes of caution and instruction as they approached, stooping once in a while for snatches of tasty river debris and tender new growth on the branches along the banks.

Twilight was almost certainly a hunting time in the late Cretaceous, as it is now, since attacking carnivores gain considerable advantage in the dim light and long shadows. Along the life-giving river, lush groves of flowering plants, ferns, conifers, club moss, and seed ferns thrived, pungent with the fragrances of blossoms from the newly arrived angiosperms and the rotting understory of the damp coastal forest. Then, as though cued by a conductor's baton, the assembling herds of dinosaurs fell silent, the only sounds the faint

DANCING TO THE FOSSIL RECORD

scufflings of small, soil-dwelling mammals in the brush and sipping sounds of the gar in the river breaking water to suck air. From my tent, I could hear only the ambient background, unequipped as I am to pick up those subtle and most dangerous signals made by the pack of dromaeosaurs downwind to the left and a lone albertosaur crashing through the brush a mile and a half away, but closing.

The attack, when it came, seemed well choreographed, a rush across the narrowest part of the river by six wolf-sized dromaeosaurs, four of which immediately fanned out to surround the nearest startled *Parasaurolophus*. Two of the attacking dinosaurs raced directly in front of the target animal, creating a diversion, hissing wildly as they ran on their muscular, ostrichlike legs past the frozen *Parasaurolophus* toward the fleeing herd. As they held her attention, the rest circled and came from her unguarded rear. The skin of the *Parasaurolophus* flared red in a display of fear and flight, and the doomed beast whistled a note of alarm akin to a screech. It bolted in the direction of the herd, but too late. The dromaeosaurs were upon her.

The first took near-fatal hits from the *Parasaurolophus*'s slashing forelegs, while two others went for her throat. By now, the fierce struggle veiled the moment in a cloud of dust, but I could still see the fourth attacking dromaeosaur avoid the slashing tail and leap to the *Parasaurolophus*'s back, like a circus monkey. Instead of tipping a funny hat to the crowd, though, the dromaeosaur began to chop with its taloned legs, digging through thick skin and flesh to the *Parasaurolophus*'s spine. The wounded attacker had withdrawn and slumped on its tail near the river, unable to continue. The others, working on the throat and belly, were now joined by the two that created the diversion, and the mortal dance ended in less than a minute. The dromaeosaurs ate well

and quickly, though they wouldn't let the wounded one near the meat. When he came near, they screamed and slashed at him. Meanwhile, the other parasaurolophids, centrasaurs, and chasmosaurs had settled like a startled flock of enormous shorebirds a quarter mile upstream from the kill. There they again bent to drink. They appeared unaware that one of their number had died, or that the giant albertosaur had used the cover of the dromaeosaur attack to take a position across the river in the forest, where it waited.

It was a long night. At dawn I got up and walked to the outhouse, feeling relieved to savor first light in the badlands and thinking everybody else must have missed the action. From the RVs and tents, I heard snores that sounded a little like snuffling *Parasaurolophi*. And from Ray's tent, the snoring sounded like an *Albertosaurus* attack. By now, Brown and Sternberg would have been up and digging, using as much of the bearable temperatures of the hot prairie day as possible. They might have heard, as I did, the common flicker rattling around in a hollow cottonwood tree, with magpies, robins, and late-staying owls getting in their two cents. A soft breeze carried the musty smell of the damp Cretaceous mud across the flats of the Red Deer River and, in passing, whispered, "Diiiiiii-innnnooooooossaaauuurrrrrrrrrrrssssssss."

HERE LIE BETSY'S BONES

DRUMHELLER, THE LITTLE TOWN where Brown, Sternberg, and everybody else began their seasons on the Red Deer River, still thrives because of the oil and gas business, a provincial prison, and the Royal Tyrrell Museum of Paleontology. The high school sports teams are named the Dinos and

DROMAEOSAURS

These fast, deadly predators, members of the *Deinonychosauria* clan— "terrible-clawed lizards"— were armed with a large, sickle-shaped claw on the second toe of each foot. They might have been warm blooded, intelligent enough to hunt in packs, and a bad dream for a grazing hadrosaur. Dromaeosaur means "running lizard."

the Dinettes. And you can't go far without running into a rock shop, a mom-and-pop museum, or a guy peddling fossils out of a pickup truck at a four-way stop sign. Half the stores, bars, and restaurants in town have neon dinosaurs in the windows, and the guy who sells you a six-pack at the 7-11 has an opinion on the Cretaceous extinction. Several sects of creationists have purposely settled in the valley to counter what they perceive as the heresies of evolution celebrated at the Tyrrell. They hold tent revivals in the summer, occupy a roadside chapel down the hill from the museum, and run an amusement park that features a giant Jesus who oversees a flock of cement dinosaurs.

When the oil boom hit the northern plains a couple of decades ago, some Alberta politicians with a little bit on the ball stuck up the petroleum corporations for a percentage of their take to build a museum for the dinosaurs. And so the Tyrrell opened in 1985. Appropriately, the superb temple to the bones of the badlands is named after Joseph Burr Tyrrell, Dawson's sidekick, who is credited with having found the *Albertosaurus* skull that started it all. In low, terraced glass and concrete buildings that mimic the stair-step terrain of the valley, the paleontologists, preparators, and curators at the Tyrrell lay out the history of life from the Archean to the present, then zoom in on the Cretaceous and the local critters. The great fossils of Alberta, which continue to pour in from the badlands, no longer end up in New York, Boston, London, and other distant museums, except by permission. If they approve of your intentions, for instance, the Tyrrell will sell you, for about $100,000, a cast of their exquisite *Albertosaurus* on display at the museum, and fossils routinely circulate on loan to paleontologists around the world. A half-million

people a year, though, show up in Drumheller to visit the Tyrrell.

More than thirty perfectly prepared monsters of the Mesozoic from around the world are on display at the Tyrrell, along with first-class fossils from every other period, all set in reproductions of their natural habitats. The cast includes *Stegosaurus, Diplodocus, Triceratops, Tyrannosaurus rex, Albertosaurus, Hadrosaurus, Chasmosaurus,* and *Dromaeosaurus.* You can climb two flights of stairs to stand eye-to-eye with *Tyrannosaurus rex,* and interpretive cards and displays tell you why, for instance, a *Stegosaurus* had those weird armor plates (they might have been a kind of radiator). At computer terminals around the museum, you can call up full reports on many of the dinosaurs, or take a very simple, utterly meaningless test to determine whether or not you are an Honorary Paleontologist. In a darkened chamber, where they are surrounded by backlit murals and bathed in the ambient sounds of their watery clicks and cackling, the fossils of plesiosaurs, mosasaurs, and ancient turtles suspended overhead come alive in the Cretaceous Western Interior Seaway.

Spectacular as the exhibits are, a honeycomb of halls, warehouses, and laboratories surrounding the

The attack, when it came, seemed well choreographed, a rush across the narrowest part of the river by six wolf-sized dromaeosaurs.

BRING EM BACK ALIVE!

When Betsy talks about another of her passions, plesiosaurs, she rows her arms through the air to imitate their swimming stroke.

OPPOSITE: Trout Waiting for Dinosaurs to Go Away

public galleries contains many times the number of specimens on view. The interior labyrinth of the Tyrrell is also home to a corps of paleontologists, who, when they are not in the field where they would rather be, are here preparing and publishing to send their discoveries into the collective wisdom of *Homo sapiens sapiens*. The profession is booming, probably because oil exploration, which once claimed most geologists, has tapered off lately, so more of them specialize in paleontology in graduate school. One of the surefire ways to find oil is to find the right color and shade of a particular minute fossil called a conodont, possibly the teeth of an elusive, primitive chordate, and one of the great puzzles of paleontology. So, shortly after the Alberta fossil rush ended, the oil rush began and a lot of earth scientists went for the money.

Not Dr. Elizabeth L. Nicholls, though, one of the blazing lights in the back shop at the Tyrrell. She's a compact, sinewy woman who looks as fit as a marathoner, talks like an auctioneer, and just loves vertebrate fossils. She began her career as a freelance dinosaur collector, like so many of the old-time fossil hunters, paying her own expenses and then trying to sell what she found to the university in Calgary. She went back to school and wrote her master's thesis on the Cretaceous turtles of Dinosaur Park and her Ph.D. dissertation on other marine turtles.

Those marvelous, tranquil animals have been aboard Planet Ocean since their emergence as land animals in the Triassic and their eventual return to the sea during the Cretaceous. Part of their tribe—chelonians—migrated back to the sea and showed up as some of the earliest treasures in the Kansas fossil war at the end of the nineteenth century. Two extinct genera—Protostega and Archelon—produced giant swimmers, some reaching lengths of over twelve feet and weighing over 6,000 pounds. By comparison, a modern leatherback, the biggest of the marine turtles, tops out at about 1,000 pounds. Remarkably, the early turtles were as highly specialized as their modern descendants, feeding on the bottom and on reefs, crunching large shellfish with powerful, horny beaks.

It's a bit of a reach to link Mary Anning and Mary Ann Mantell, who enjoyed no official standing, with Betsy Nicholls, who ascended to the top of a profession that was exclusively dominated by men until just a few years ago. Her cubicle in one of the labs is spartan, like those of most paleontologists for whom life in the field is real and all else is preparation and annoyance. Over her desk, somebody has tacked a postcard of a man in the classical thinking pose, elbow on knee, chin on fist, with the legend "Study the Past" chiseled in the block of stone on which he sits. Taped to the top drawer of one of her gray specimen cabinets is a label, "Here Lie Betsy's Bones," under a picture of a mosasaur skull. Among her many achievements, Betsy described the first *Hainosaurus* found in North America, a fifty-foot mosasaur and the biggest of all the marine reptiles. Her conversation dashes from critter to critter, all Mesozoic vertebrates, most of them marine reptiles. When she talks about another of her passions, plesiosaurs, she rows her arms through the air to imitate their swimming stroke. "They swam the way turtles do," she says. The animation in her eyes reflects a clear vision of plesiosaur mechanics, which has been a topic of some controversy.

Betsy says paleontology is thriving as a profession and a hobby because fossils are being popularized, and because we are finally getting around to dismissing the limitations of anthropocentric creationism. "We want to be able to solve all our

TROUT WAITING FOR DINOSAURS TO GO AWAY...

problems and be the center of the universe, but the fossils tell us we're just another one of the pieces," she says. "I consider myself to be a part of the world and a very small part of it at that. I'm an active environmentalist because we simply have to share what's here and leave as much as we can for whatever's to come. We're just not sure what that is, and you have to be able to live with that uncertainty."

One of her colleagues at the Tyrrell, paleo-ichthyologist Andy Neuman, who's working mostly on Triassic fish, takes a slightly different approach to the thorny confrontations with the local creationists. He was a science teacher in Edmonton before landing his job as a curator at the museum and is an attentive generalist when it comes to the story of life. "We have people coming through the museum all the time who complain bitterly about the lack of space given to creationism. We don't celebrate it, but nowhere in this place do we slam it," he says. "It's really only the concept of a beginning that is a problem, since we can interpret time metaphorically. And it's difficult for me, and a lot of good scientists, to believe that life on earth is just random."

His office, like Betsy's, is typically not homey or decorated: a couple of drawings of his current critters, a Triassic fish with a vague resemblence to gar and a small marine reptile something like an iguana, some clutter of books, and a few cartoons taped to available surfaces. His worn, dusty field boots are parked next to a filing cabinet and give the lie to everything indoors about the man.

"Anthropocentrism is the great flaw in religion and creationism," he goes on, echoing Betsy Nicholls. "The great lesson of paleontology is that we are not at the center of things. We are not better environmentalists because we think earth belongs to us and that we can fix anything. We are not in control." Andy says topics like the social and environmental responsibilities of paleontologists come up over beer and pretzels at scientific conventions. "We can destroy ourselves, there's no doubt about that, but we really can't take the planet with us as some like to imply. Maybe a few animals and plants, but not the whole earth. We still must be responsible, even if total destruction is not the outcome. We simply must consume less," he says. "Balance is the key, and nothing speaks of balance like holding the fossil of an extinct animal from 65 million years ago in your hand."

DINOSAURS PHONE HOME

THE SOUL OF PALEONTOLOGY is speculation, but you have to watch your step in the vicinity of fantasy. Among the denizens backstage at the Tyrrell, standing alone in a vacant office in a storeroom, is a replica of Dale Russell's dinosauroid. Russell hypothesized that if the Cretaceous extinction had not killed off the dinosaurs and some had continued to evolve, a good candidate for intelligence would have been *Stenonychosaurus inequalus*, a late Cretaceous carnivore he found in the badlands in 1968. Like *Dromaeosaurus*, *Deinonychus*, and *Albertosaurus*, this animal was bipedal, or walked on two legs, stood three feet tall, had an almost-opposable thumb on its foreclaws, possibly created a single image from its two big eyes, and, most critically, had a big brain for a dinosaur. Had *Stenonychosaurus* survived, Russell mused, what would its twentieth-century descendants have looked like? His fiberglass conceptual model, the original of which is in the National Museum in Ottawa, looks an awful lot like the Stephen Spielberg box-office space voyager, E.T. Naturally, dinosauroid got a lot of ink in the mass

"As a child, I could imagine the world of dinosaurs. At times I was a dinosaur. And so rather than do something practical when I grew up, I just stayed with dinosaurs."

—DALE A. RUSSELL
In an interview with
John Noble Wilford from
The Riddle of the Dinosaur

media, which always makes paleontologists suspicious. Nothing can make you more suspect to your colleagues and competitors than getting your name or picture in *Time* magazine.

But no dinosaurs survived, nor did any of the great Mesozoic marine reptiles. Had any made it through the Cretaceous extinction, at least some evidence would have shown up in the fossil record of the last 65 million years. Not one shred has appeared. The recent theory that birds are really dinosaurs is already falling apart, no matter what the popular magazines say, though few paleontologists disagree that birds and dinosaurs had a common reptile ancestor at some point. Too bad robins aren't dinosaurs. And the next time somebody tells you a Japanese fishing trawler has landed a plesiosaur, or that the Loch Ness monster is really an *Elasmosaurus* that survived, get down a bet against and make some money.

DIG HARD, SELL HIGH

HERE'S A TYPICAL STORY. At the 1991 Denver Rock and Fossil Show, a collector from Calgary spotted a rare fish fossil, *Eosalmo driftwoodensis*, a 50-million-year-old ancestor of modern salmon and trout. Recognizing the enormous value of such a fossil to a museum, he called the Royal Tyrrell, brokered a deal so he'd make a percentage of the price, agreed to deliver it to Drumheller, and bought the fish for $5,000. So far, so good. At the border, though, the collector was busted because he didn't have a permit for the fossil's original exportation from Canada, a requirement under a recent law. Even though the fossil left the country in the hands of an American collector long before the law was passed, it was contraband in Canada and technically

"YOU GOT A PERMIT FOR THOSE TRILOBITES SON?"

nobody owned it, could buy it, or could sell it. Eventually, in a hard-won compromise, the Calgary collector donated it to a museum in Quebec.

Here's another one. On May 15, 1992, the FBI raided the Black Hills Institute in Hill City, South Dakota, and seized the biggest and best specimen of *Tyrannosaurus rex* ever found. The treasure is nicknamed "Sue" after Susan Hendrickson, who, in 1990 on a walk along a butte with her dog, Gypsy, noticed a fragment of backbone. (Another woman, another walk, another great discovery.) Hendrickson is not part of the ownership dispute—but not to worry, plenty of other people are. Hence the FBI's raid on behalf of one of those would-be owners, the People of the United States. Hendrickson, who happened to be a skilled field scientist, was also close to the founders of the Black Hills Institute, a private paleontological company that not only digs and donates fossils to its own museum but sells them to collectors around the world. That's two potential owners, the United States and the Black Hills Institute. But there are more. The *T-rex* was found on land leased by one Maurice Williams, who accepted $5,000 from Black Hills Institute president Peter Larson for permission to remove the fossil. That's three. Larson said he intended to donate the *T-rex* to his foundation-owned, non-

profit Black Hills Museum. Then the Cheyenne River Sioux, deeded owners of the leased land, got into the act, and, finally, so did the federal government, in general, because it claimed ultimate title according to Bureau of Indian Affairs deeds of trust.

Why all the brouhaha over who owns the bones? On the open market, the biggest, best *Tyrannosaurus rex* on Planet Ocean is worth several million dollars.

Different governments have different rules for fossils found within their jurisdictions, so you can never be sure what's what unless you ask. Most of the pressure to keep fossils in the public domain, as in Canada, comes from paleontologists and museums, who are afraid somebody could discover a critical link, say, a major transitional animal, and sell it to a private collector who might never breathe a word about it to the scientific community. Commercial fossil hunters say science should just get into the market, that they trade and sell fossils among themselves all the time anyway. Big bucks are a real issue, since many commercial collectors have legitimate permits to collect in one place and so can market fossils taken from other places illegally. A flourishing black market surrounds every rock and fossil show in the world, and if you know what to ask for, you can probably buy it. Cope, Marsh, and the other old-time fossil hunters were more like gold rush prospectors than even they supposed. If they were alive now, wheeling and dealing with their archelons, mosasaurs, and dinosaurs, they would have piled up enormous fortunes.

NOT IN KANSAS ANYMORE

1. STYXOSAURUS
2. BANANOGAMIUS
3. THALASSOMEDON
4. BRACHAUCHENIUS
5. ARCHELON
6. SAURODON
7. PLATYCERUS WITH OMOSOMA
8. CRETOXYRHINA
9. SQUID
10. GIANT COELACANTH
11. BANANOGAMIUS
12. CLIDASTES
13. PACHYRHIZODUS
14. PROTOSPHYRAENA
15. TYLOSAURUS
16. PLATECARPUS
17. TRINACROMERUM
18. PTERANODONS
19. HESPERORNIS
20. XIPHACTINUS (SWORD RAY)

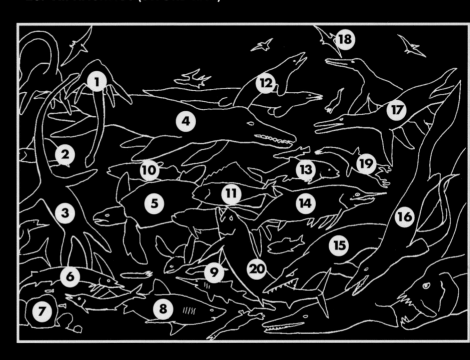

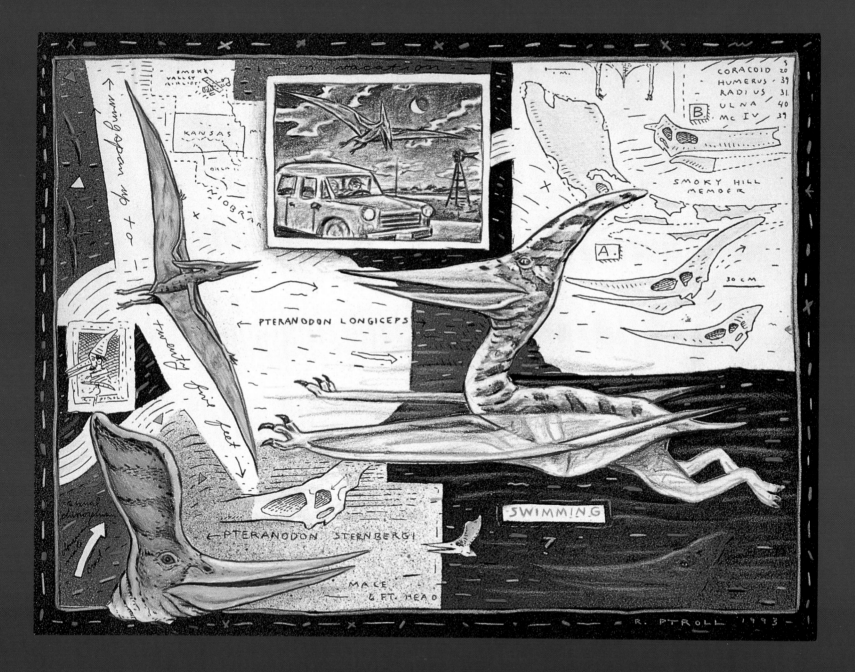

Swimming Pteranodons

WHEN YOU WISH UPON A GAR

KANSAS CRETACEOUS

GAR, BULLETS, AND DISCOUNT BEER

IT WAS WAY TOO EARLY on an already hot June morning to be jammed onto the bench seat of a pickup truck headed south out of Wichita, swapping gar puns. But wherever you go, there you gar. And there we were with Steve Fairchild, the Guru of Gar, flashing past the suburbs and the air base on the way to a sandpit for a little fishing on the Arkansas (when you're in Kansas, that's Ar-Kansas, not Ar-Kensaw) River. During the silences that followed particularly bad puns—When You Wish Upon a Gar; there goes a Gar-V—the highway took over and chanted "Kan-sas, Kan-sas" as our wheels strummed the seams in the lime-white concrete. Steve was a pal of Ray's from high school. While the race riots of the early seventies made life miserable, they were learning to be artists and musicians and trying to keep their heads down, like Cretaceous mammals hiding from dinosaurs. Steve hung on where he was because Wichita is "off the hip track," as he says, and now he works a day job running a chain restaurant called the Olive Garden. Ray went west but was back in Wichita for his twentieth high school reunion.

I was there for my first real dose of this fertile patch of Planet Ocean. I'd never seen so many Dairy Queens and well-fed citizens, and the wheat absolutely sang of abundance from the ripe, yellow fields. Wheat was a hot topic, too, because the spring rains weren't letting up on schedule and you can't harvest wet wheat. At the time, the farm belt was like a big casino: the longer the rains lasted, the less the wheat would be worth. So a lot of farmers were buying and selling the crop to each other or to commodities speculators at future prices to hedge against losses in the fields. Even if the wheat was worthless, smart traders would survive.

Steve Fairchild wasn't all that interested in

"Man is so constituted that he considers the value of other living things solely on the basis of his own comfort and convenience. On this basis, much has been said against the gars and very little in their favor."

—A. C. WEED
The Alligator Gar (1923)

VELOCIRAPTORS VS. GAR

*A ten-pound gar is three feet long
and four inches in diameter, an
ambush predator with the patience
of a sunning snake and the scare-me
jaws of needle-sharp teeth a lot
like those of an ichthyosaur.*

wheat futures, or even in the fossils of the legendary
Kansas chalk that Ray and I had come to see. But
he got deeply into gars in the early eighties because
he wanted to avoid the metal-flake, bass-boat mad-
ness that infects everybody else in Kansas. Bass
fishermen hate gars. When they catch one, they torture and maim it, then maybe nail the head and skin
to the chicken house door. Gars are the underfish.

In many places, entire watersheds are poisoned to get rid of the gars, and some states have a bounty on them. "I wrote to the fish and game department," Steve said, "and asked which rivers have gar in them and they said, in so many words, 'Every damn river in Kansas, you fool.'"

But Steve found the town of Virgil, a little backwater a couple of hours from Wichita with the appropriately seedy Clifton Hotel near the river, and there he organized an annual gar tournament from 1987 to 1989. Most of the fishing was done from barstools. The contestants ran their lines out the door of the Clifton's saloon and down the street to the river and drank substantial amounts of the cheapest beer they could buy. According to Steve, you can't catch gar if you drink premium or imported beer. They sang songs like "Mama, Don't Let Your Babies Grow up to be Gar Fishermen" and told gar stories—"I saw one bite my father so bad we had to quit fishing, and it had been lying in its own blood in the bottom of the boat for an hour." Not a single gar was ever caught during the tournament, but Steve awarded prizes anyway, one year including a customized Gar Car, which made the newspapers. "My wife won't let me tell gar stories in the house anymore," Steve says. "She lost her sense of humor about gar the night a thunderstorm tore up our campsite during the last tournament."

The bacchanals in Virgil were a sendup of the million-dollar bass fishing tournaments, but they could just as well have been a sincere tribute to the guest of honor, one of the most accomplished survivors in the story of life. A ten-pound gar, for instance, is three feet long and four inches in diameter, an ambush predator with the patience of a sunning snake and the scare-me jaws of needle-sharp teeth a lot like those of an ichthyosaur.

Gar—*Lepidosteus*—probably appeared sometime during the last few million years of the age of the dinosaurs at about the same time as paddlefish, a distant relative whose rare descendants still live in the rivers of Kansas. Gar are thriving, though, in four distinct species: spotted gar, short-nose gar, long-nose gar, and the giant alligator gar, which may reach lengths of twelve feet or more. A big one in an aquarium tank will sure hold your attention. The gar's respiratory system can absorb oxygen from both water and air, a slick adaptation for survival in warm, slow-moving water with low levels of oxygen.

Steve had to go to work that afternoon, so we didn't make it all the way to Virgil. He'd heard there were gar in the sandpit on the Arkansas where it flows through a farm town that announced its philosophy on a sign as we drove in past the Dairy Queen: "Welcome to Derby, where people care about each other." The crew at the sandpit office *were* helpful and sent one of their dump truck drivers to lead us through the mountains of dredged sand to the river. Scotty was a big man, with an honest-to-god crewcut, a twenty-inch neck, and arms the size of holiday pot roasts. He wasn't at all put off by three guys looking for gar. "Yawl like gar, huh? Me too. Always have. Bass don't do nothing for me. They just aren't much excitement. Gar, though, they'll scare you. They look right at you. They know you're there. I've heard of bullets bouncing off of 'em."

Midmorning was T-shirt hot and muggy, with a nasty hint of a sweltering, heart-of-America day to come, when even a rainstorm would be just another setting on the bake cycle. Outsiders have been known to beg shamelessly for air-conditioning. The Arkansas at Derby is a modest remnant of the mile-

LAND AND SEA DURING THE CRETACEOUS
65 to 140 million years BP

The headline event of the Cretaceous was the emergence of flowering plants, the angiosperms, which would eventually blanket the world with plant life. Dinosaurs were still boss on land, and the great marine reptiles enjoyed their golden age in the sea, feeding on abundant ammonites, belemnites, fish, and each other. Everybody knows what happened next. Snake eyes.

OPPOSITE: *Velociraptors* **vs. Gar**

The gar rolled like shiny logs against the brown Arkansas and punched the surface with their long snouts to sip at the air, just as they did 65 million years ago.

wide torrent it was before the state of Colorado built a dam upstream at Trinidad in the early eighties. Now, most of the river's flow comes from what had been just a tributary, the Little Arkansas, which carries the debris of its passage through Wichita to the north. Plastic and other floating garbage was piled in a frothy scum against the pipelines and stern of the sand dredge anchored in the river. As we approached, Scotty launched into a diatribe about Wichita. "It's this kind of stuff that makes me want to be an environmentalist," he said, not entirely convinced. "There's just too damn many people." Everybody cracked a beer except Scotty, who left to deliver a load of sand. The politics of water in the next century are going to make the politics of everything else seem like a cheerful parlor game.

We rigged our fishing poles with the snarled ball of monofilament line and red cloth you use to snag gar. You don't use hooks, because they won't penetrate the hard, scaly skin of the gar's mouth, and the tight rows of razor-sharp teeth will tangle up with anything the fish gets in its mouth anyway. We cast for a while, but the river was too low, or too hot, or too something, and we didn't catch any gar. We did, though, see them roll like shiny logs against the brown Arkansas and punch the surface with their long snouts to sip at the air, just as they did 65 million years ago. Then, instead of being a bunch of guys standing around drinking cheap beer on the riverbank, we might have been a herd of *Triceratops*, or a gang of *Deinonychus* with meat in mind. Or, later, we could have been some of those giant birds that dominated the land food web in the early epochs of the Cenozoic, the New Age. And still later, we might have been a band of Neanderthal primates, hunkered down in the heat of the day, listening to the sucking gar and waiting for an

animal with its guard down to show up for our next meal—unaware, of course, that our own extinction was just around the bend. And all the time, the gar have watched from the river.

Scotty came back from his delivery and seemed to feel personally betrayed by the Arkansas River that we hadn't caught anything. He insisted on taking us out to a buddy's mobile home in the country to show us some gar skulls and skins tacked to the bleached wall of his shed. The heads were dead ringers for those of other marine creatures, alive or extinct, that made a living by getting in a good first bite on something slippery. Some marine hunters, like bass, are built differently. They learned to suck as well as bite, turning their heads into vacuum cleaners and forming bodies that specialize in speedy pursuit instead of the sudden lunges of gar and other ambush predators. On the way back to Wichita, we stopped at a pond known for its gar, unable to let go of our day with this tough, wonderful character, this ancient fish. Again, our casting was futile, but where we parked the road was littered with gar skins and skeletons, part of the owner's campaign to cleanse his pond with chemicals so he could plant bass. Ray and I found a couple of nice skulls, and when we hit the road a couple of days later for the fossil beds of northwest Kansas, we duct-taped one to our front bumper. For luck and for survival.

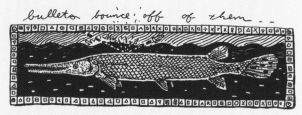

The whole extravaganza is festooned with angels, devils, the Goddess of Liberty, and a red, white, and blue American flag.

BACK TO THE GARDEN OF EDEN

IF YOU DRIVE sixty-five miles northwest from Wichita on Route 81, you come into the Cretaceous in the upper left-hand corner of McPherson County, right around Lindsborg, where, coincidentally, young Raymond Troll went to college. But who really believes in coincidence? There we were, on a pilgrimage to the fossils of the Kansas chalk twenty years after Ray just happened to have worked for J. Harold Caldwell drawing educational filmstrips (remember those?) on the geology and fossils of Kansas, and where he also just happened to follow his art habit to Bethany College in Lindsborg, which just happens to be tucked into the southeastern edge of the Smoky Hill River valley, which just happens to expose the most famous fossil beds in the world. Anyway, we stopped in Lindsborg for a sandwich at the Old Stuga, a dark but friendly bar that has helped generations of students and professors depart reality and where the owners have varnished layers of photographs of their customers into the tops of big round tables. The pictures are a big hit, probably because being in the table strata imparts a measure of immortality, tan-

gible proof that you've passed by, evidence that will outlast your physical presence. We found Ray's picture in one of the tables, talked with the regulars about how the place had changed (but not the good food), ate our turkey sandwiches, took a long time saying good-bye, and headed for the Cretaceous.

Kansas is a stratigrapher's dream. The tame grids of square counties make it easy to compare the road map to the Generalized Geologic Map and see that on the way from Wichita we nicked a sliver of Permian outcropping, traversed some recent Quaternary rock, and, just south of Lindsborg, skirted a wrinkle of Tertiary finally to reach the Cretaceous chalk. The surface terrain signals the transitions between geologic ages and, not incidentally, told several waves of human settlers where to find water, salt, oil, and, during the glory years of the Marsh and Cope bone boom, fossils. All of Jewell, Republic, Osborne, Mitchell, Cloud, Russell, Lincoln, and Ottawa counties and parts of Wallace, Logan, Gove, Trego, Ness, Hodgeman, Ford, Phillips, Rooks, Ellis, Rush, Pawnee, Smith, Barton, Ellsworth, Rice, Saline, McPherson, Marion, Washington, Clay, Clark, Comanche, and Kiowa counties occupy a contiguous outcropping that was part of the basin of a great, shallow inland ocean during the Cretaceous.

The interior sea formed and dried eight or nine times during the 80 million years of the Cretaceous, probably because Planet Ocean was in a particularly violent tectonic phase. Maybe the moon was wobbling, or some dark mass from the cosmos paid a visit, or the sun's magnetic halo simply tipped Planet Ocean on its side. Everybody has a theory. Whatever, the single landmass of the Triassic and Jurassic

was breaking up into the continents as we know them today and crust building in the oceanic rifts was at an all-time high. The additional land displaced seawater from the ocean basins, forcing it into the central lowlands of the newly forming continents, including North America, where as much as a third of what is presently land was submerged.

Paleontologists call that come-again ocean the Western Interior Seaway, the Kansas-Nebraska Sea, or, in this particular formation, the Niobrara Sea after a county in Nebraska where the Cretaceous outcropping was first described and named. The place was loaded with marine reptiles, flying reptiles, birds, fish, and invertebrates, though most of the spineless tribe had calciferous shells and didn't make it because the chemistry of the Niobrara destroyed their remains before they could become fossils. By the middle of the nineteenth century, the Niobrara, by then Kansas, was also loaded with farmers, ranchers, preachers, soldiers, Indians, fossil hunters, and vertebrates of every stripe looking for the land and opportunity promised by Manifest Destiny, a curious vision of fate that told European settlers that they owned everything they found. Many were a few sandwiches short of a picnic, to put it politely, setting up on the way-and-gone prairie to live out fearful, hopeful, independent, or simply distorted realities. S. P. Dinsmoor, for one, was a retired farmer and Civil War veteran who spent his later years until his death in 1932 sculpting a vision of creation out of limestone logs and concrete trees. He called it, of course, the Garden of Eden. This strange and delightful remnant of one man's story of life has been preserved by the 600

townsfolk of Lucas, who enjoy a modest flow of tourist money from it, and by a fund-raising campaign by Kansas artists who thought the loss of such a great collection of folk art was unthinkable.

So we drove to Lucas and argued most of the way about a particular kind of giant short-necked plesiosaur, *Brachauchenius lucasi*, whose jaws alone were six feet long. Usually, whoever wasn't driving read aloud from the pile of books we accumulated every time we went anywhere. We had a critter of the day, and the reader was supposed to survey the library for everything on, say, plesiosaurs. The reading didn't last long after we left Lindsborg, because I got Ray going on *Brachauchenius lucasi*, just about convincing him that maybe this obscure specimen of a marine reptile the size of a long-haul semi, the skull of which is now in the Sternberg Museum in Fort Hays, was found right there in Lucas, hence the name. He was pretty excited by the prospect of doing some real paleodetective work. Since he'd been to the Garden of Eden many times before, Ray wasn't as enthusiastic as I was about the fifty-mile detour from our beeline to Keystone, where we were actually going to go fossil hunting, but my *Brachauchenius* ploy worked. So while I took the guided tour of the Garden of Eden, Ray went looking for truth and rumor about *Brachauchenius lucasi*, and for Brant's Market, famous for homemade bologna.

The first thing you have to understand about the tour of S. P. Dinsmoor's quirky vision of creation is that everybody who's read the brochure knows the climax, which inspires a certain amount of dramatic tension. In one corner of the block-square garden, Dinsmoor built his own mausoleum, and there he lies, in a glass-covered coffin. Not many people have really seen a lot of dead bodies. Before you get to the crypt, a slow-walking, fast-talking guide leads

SAURODON

PROTOSPHYRAENA

IN DINSMOOR'S CRYPT

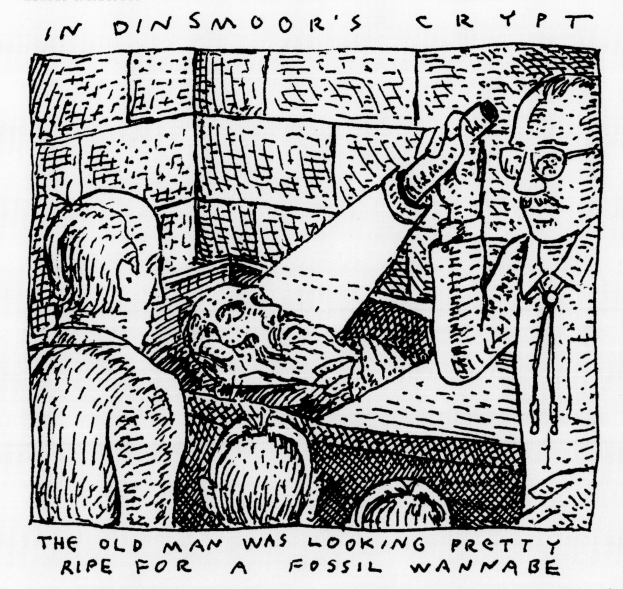

The tomb was cool inside, insulated from the Kansas heat by foot-thick limestone logs, and pitch dark with the door closed behind us.

THE OLD MAN WAS LOOKING PRETTY RIPE FOR A FOSSIL WANNABE

you through the central house, built of rock logs cut from the Kansas chalk, and shows off Dinsmoor's exotic chessboards, his barbed wire collection, and the beautiful finish work and moulding that apparently fascinated its eccentric carpenter. Outside, you begin with the beginning according to Dinsmoor. He fashioned his intricate sculpture from 113 tons of cement, forming a high fence of characters and drama around three sides of the house and garden.

The story line is a little bit along the lines of "I Know an Old Lady Who Swallowed a Fly," a statement of death and dependence that says somebody or something is chasing all living things all the time.

Life-sized, loin-clothed Adam and Eve are the first link in the chain of evolution, which connects a descending hierarchy of living beings, according to Dinsmoor. On the twisted concrete branches and struts, he hung the figures of a soldier chasing an

PLEASURES OF THE PLESIOSAUR

Pleasures of the
Plesiosaur

Indian, who in turn is after a dog, and the dog is after a fox, and the fox a bird, and the bird a worm, and the worm is eating a leaf. The whole extravaganza is festooned with angels, devils, the Goddess of Liberty, and a red, white, and blue American flag, the first concrete flag in the world, lit at night by a series of concrete lamp standards that project at odd angles from the structure. "Mr. Dinsmoor was proud of electricity," the guide notes. Next to the mausoleum Dinsmoor created a massive triptych he called "The Crucifixion of Labor," depicting the death of the workingman under the lash of lawyers and bankers.

Finally, the ultimate moment arrived. With a rattling ring of keys, the guide opened the door to the crypt and handed me a flashlight. "This is a tomb. Mr. Dinsmoor is actually in here, and you'll be able to see him," he chanted. "Please show proper respect and take no photographs." The tomb was cool inside, insulated from the Kansas heat by foot-thick limestone logs, and pitch dark with the door closed behind us. There, washed by the pale beam of my flashlight, in a cement coffin he built himself, was the corpse of S. P. Dinsmoor, sixty years after he drew his last breath. His body had shrunk to half its size in life, the skin was drawn tightly across the bones in his face, and his hair, teeth, and chin whiskers were exaggerated by the retreating flesh. I heard the guide breathing, standing right behind me like he was waiting for me to faint—which no doubt a few of his customers have. But I was thinking, well, in a million years, a good mud flow could cover Dinsmoor and shut off the oxygen that will otherwise turn his bones to dust. He'll make a beautiful fossil laid out like this. Paleontologists or whoever's around to care in the Neo-Cenozoic, after the next mass extinction, will no doubt be confused by the rubble of the Garden of

Eden collapsed around this nicely articulated vertebrate. Maybe solving the puzzle will unify a school of thought, or kindle a fierce rivalry of interpretation, or start a fossil war. Or maybe the fossilized Dinsmoor will remain undisturbed and emerge only eons later as a reef-building wrinkle in some new-grown sea bottom, ruled again by invertebrates that inherited the genetic force that will shape creatures for another future on Planet Ocean.

"Time's up, Sir," the guide said after just a couple of minutes, opening the door to flood the crypt again with the white light of midday Kansas. "Got to move on."

LIFE WITH FATHER

RAY DIDN'T MAKE ME pay too much for sandbagging him about *Brachauchenius* to detour through the Garden of Eden. He even shared his Brant's bologna with me as we blew across the prairie on the way to Keystone to hunt bones with the Bonners. We took turns driving and reading up on plesiosaurs, then stopped at the Sternberg Museum in Hays to see the actual *Brachauchenius lucasi* skull, collected, it turns out, on a ranch in Russell County in 1951. The place is small and dusty, nothing like the Royal Tyrrell, the Smithsonian, or the other grand museums. But the specimens are terrific, and most were collected from the Kansas chalk within a hundred miles of Hays.

The first, most celebrated fish-within-a-fish, for instance, is on the wall to the left as you walk into the hall of vertebrates. George Sternberg found this one; Marion Bonner found the second, which is in the Royal Tyrrell. Originals and casts of others are in museums all over the world, because the fossil tells a revealing story about life in the Cretaceous sea. This fourteen-foot *Xiphactinus*—"sword ray" —

JUST ANOTHER VERTEBRATE

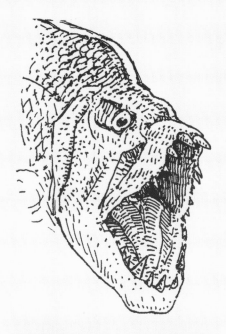

HESPERORNIS

Both *Hesperornis*, the "Dawn Bird," and its smaller cousin *Baptornis* were big marine birds of the Mesozoic. Their beaks were lined with teeth, and so were perfectly equipped —like gar and ichthyosaurs—for dining on slippery critters. *Hesperornis* was flightless, though a true bird that looked a little like a giant modern loon. Some were six feet long. Their fossils are rare, probably because they settled slowly to the sea bottom after death, giving other predators a shot at an easy meal.

had, it seems, been dining on its six-foot cousin, *Gillicus*, a terrible moment for the smaller fish. *Xiphactinus* and similar predators routinely swallowed their prey headfirst and whole. Because the tail of a fish like *Gillicus* was a dangerously sharp collection of segmented rays, *Xiphactinus* snapped it off with its enormous jaws on the way down. Apparently, though, many an aggressive diner failed to bite off the tail of its last meal, which, squirming in the gullet on the way down, pierced the buoyancy sac of the soon-to-be-dead predator. Both went to the bottom and the fossil-making mud of the Niobrara Sea. In the Mesozoic, a fish eating a fish was the equivalent of cigarette smoking: you know it's fatal a lot of the time, but you just have to do it. The result, there on the wall at the Sternberg Museum, is a world-famous crowd pleaser, because it describes not only a representative animal of the Cretaceous but also a dramatic, clearly defined moment in time.

Next to the ancient and mortal pas de deux between *Xiphactinus* and *Gillicus* is the Memorial Mosasaur, *Tylosaurus proriger*, collected by George Sternberg and paid for and donated to the museum, according to a brass plaque, by the Fort Hays State University seniors of 1927, 1928, and 1929. The mosasaur was their class gift. George found the thirty-foot sea dragon seven miles southeast of Russell Springs. Around the corner, in a glass case, are the world's best specimens of *Pteranodon walkeri* and *Pteranodon sternbergi*, the two species of the great pterosaurs—flying reptiles—that filled the skies over the Niobrara Sea. Right next to them is what most paleontologists consider to be the best specimen of a short-necked plesiosaur in the world, a nearly perfect twelve-foot *Trinacromerum*. Marion Bonner found this one and donated it to the museum. "We were out in the field, all of us, my late wife and

most of the kids," Marion Bonner once recalled in a television interview, "and I took one look at this *Trinacromerum* in the chalk and said, 'Margaret, get the kids out of here, everybody get going.' I could tell it was perfect, with the head and everything intact except one paddle, and I didn't want to take any chance that it could get damaged. Later, she said, 'Marion, I'm surprised at you. You disowned us for a plesiosaur.'" Afterward, she and Marion painstakingly prepared the perfect plesiosaur together.

Across from Marion's plesiosaur the enormous skull of *Brachauchenius lucasi* dominated the wall, the business end of an animal forty feet long and hungry all the time. And sure enough, it was found in Russell County. Ray was smug as we walked out into the late afternoon, talking about what a fine time Sternberg, Bonner, and the rest of the early paleontologists must have had in the Kansas chalk, being the first to find these astonishing sea lizards.

The wounds of the fossil war between O.C. Marsh and Edward Drinker Cope have healed completely in the 120 years since they joined the fray, but you can still trace their lineages through several generations of Kansas collectors. Most who worked with Marsh came to despise the man. Nonetheless, many of his apprentices and successors became giants in the field, including B.F. Mudge, Samuel W. Williston, and Barnum Brown. Williston took his first collecting job with Marsh the summer before he entered medical school in Iowa, graduated, followed Marsh to Yale College, where he earned his doctorate in paleontology, then returned to Kansas to become a professor at the university at Lawrence. According to most reports, E.D. Cope was not an entirely pleasant man either, but he spawned Charles H. Sternberg, the patriarch of a family of fossil hunters that included his twin brother, Edward, and

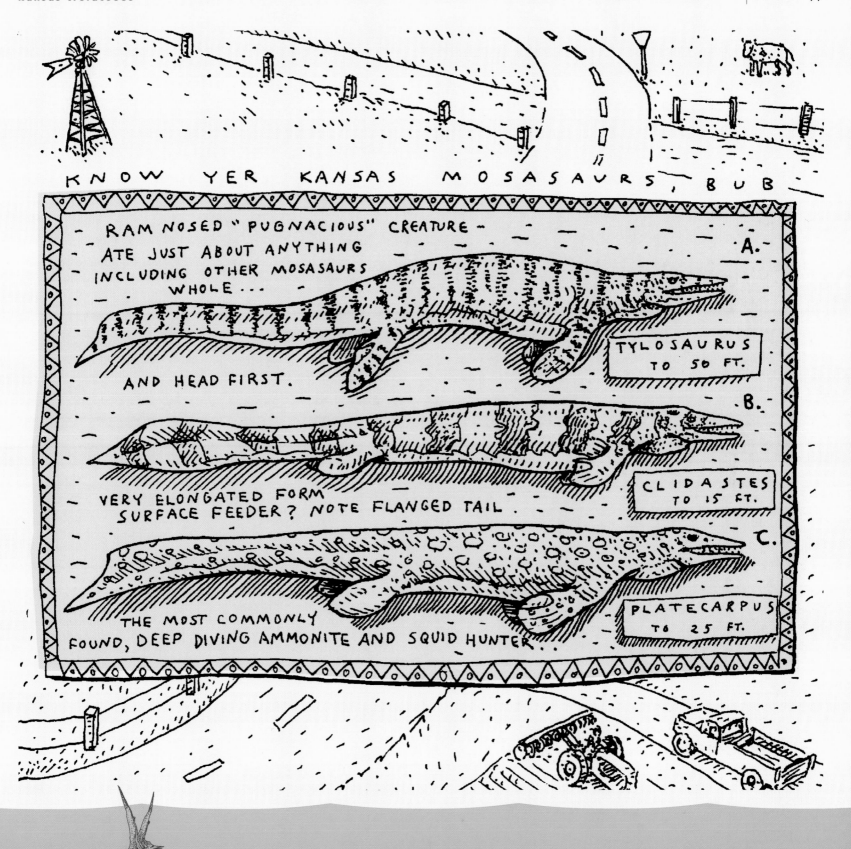

THE GREAT WESTERN INTERIOR SEAWAY OF NORTH AMERICA

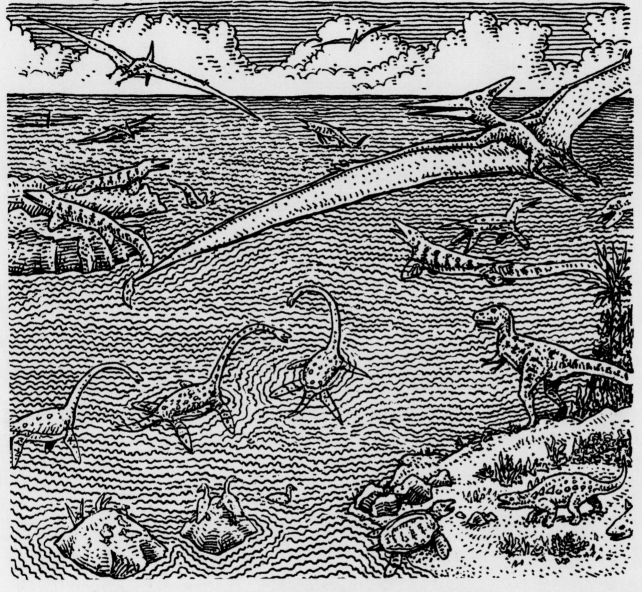

During his eighty-something years, Marion painted houses, ran the picture show in Leoti, Kansas (pop. 1,800), fathered eight children with his wife, Margaret, and spent every spare hour in the chalk, often bringing his family with him on extended sojourns. He found the fossils of just about every kind of plant and critter in the Niobrara Sea. Like other freelance collectors, he sold some, gave some away, and sometimes worked on contract to a specific museum. Marion had a long-standing relationship, for instance, with the Natural History Museum of Los Angeles County, which owns many of his best fossils. Marion's oldest son, Orville, studied with George Sternberg at Fort Hays State and became a collector and preparator at the University of Kansas, where, over his desk in the museum basement, he has a sixteen-by-twenty-inch, framed photograph of E. D. Cope. Another of

OPPOSITE: Professor Williston's Last Supper

his sons, Charles M., Levi, and George. George became a professor of paleontology at Fort Hays State College, where his premier freelance collector was Marion Bonner, who died, coincidentally, the week Ray and I arrived in Kansas on our tour of Planet Ocean.

Marion Bonner's sons, his sixth child, Chuck, is an old friend of a friend of Ray's, an artist, musician, and freelance paleontologist himself who lives at Keystone, smack in the middle of the legendary beds south of Gove. We were drawn to the place as though on rails.

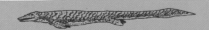

THE CHURCH OF THE CRETACEOUS

The Niobrara was rarely deeper than 900 feet, which by ocean standards is a kitchen sink.

OPPOSITE: **The Meanest Pescado**

DRIVING ACROSS KANSAS is Ommmmmm on the range, a mantra set to the sizzle of tire noise and interrupted, once in a while, by the need to turn the steering wheel a degree or two. As evening caught up with us after Hays, the critter of the day was the mosasaur, but Ray nodded off reading a description of gastroliths, the stones some paleontologists think the giant lizards swallowed to help them mulch their food, just like a lot of modern animals. Fossils have been found containing rocks that only could have formed hundreds of miles away from where the mosasaur died. That means the beasts might have migrated or roamed all over the Niobrara. So Ray was sleeping, and I was trying to imagine a swarm of mosasaurs headed south, like we were, swimming fast instead of hunting and feeding. Maybe they were on their way to squeeze out their spring crop of young in warmer climes, or maybe they were fleeing some catastrophe up north. Most likely, mosasaurs weren't real social, preferring solitude except when breeding, and without a doubt never crossed species lines except in my mind.

But just after we crossed the Smoky Hill River, I caught the shimmer of water out of the corner of my left eye. Suddenly, a flotilla of giant sea lizards was pacing us on the prairie-turned-to-ocean. Ten-foot-long *Clidastes*, the sports cars of mosasaurs, were juking ahead of the bigger, slower fifteen-foot

Platecarpus and the thirty-foot *Tylosaurus*—all propelled by their powerful, slashing tails, like gigantic crocodiles. Some dove and surfaced, their writhing bodies slapping the water, and they snapped at each other when they got too close but never pressed an attack past the first irritable nip. I heard them hiss and whistle, like snakes the size of Lear jets, and thought for just a moment that I smelled their rank collective breath.

In the 500 million years since the Cambrian explosion, the creatures of the sea had sorted themselves roughly into the streaks of life we know today. Birds, fish, reptiles that fly and swim, invertebrates from sophisticated ammonites to worms, and plants were all there. Only mammals were missing. Still ashore, members of our tribe were scuffling around in the brush trying to avoid dinosaurs and would not colonize the oceans until after the Cretaceous extinction. What the hell, I thought, if time travel is a dream anyway, I might as well have scuba gear with me. So I left Ray asleep in our speeding truck, stood on the running board to slip into my wet suit, and, figuring I would be invisible because mammals wouldn't return to sea for 5 or 10 million years, I took a swim.

I eased from the highway across a shoal of oysters and flat clams, backing in so my fins wouldn't snag and trip me, testing my regulator and spitting into my mask to keep it from fogging. The ocean itself felt surprisingly familiar. The sounds of surf on an offshore reef carried across the shallows, and the radiant shades of color from aquamarine to cobalt blue marked depth and bottom contours. At the tide line, the sea ended and the land began with waves arriving and withdrawing in currents, rips, and eddies, much as it does now. The water of the Cretaceous sea flowing around my legs was warmer than the Caribbean, and I knew the wet suit would

XIPHACTINUS

A TELEOST FISH OF THE LATE CRETACEOUS — WORLDWIDE DISTRIBUTION. POSSIBLY THE MEANEST PESCADO OF ALL TIME

CRETACEOUS ROAD DREAM

Cretaceous
Road Dream

be too hot until I cleared the reef. I remembered how relatively shallow the Niobrara was, rarely deeper than 900 feet, which by ocean standards is a kitchen sink. The sun warmed and transformed this salty sea through nine or more cycles from dry to wet over millions of years, creating and destroying marine habitat, a rhythm that was perfect for rapid evolution. My suit *was* way too hot, but it afforded me at least the illusion of protection that my bare skin would not. The mosasaurs were out tearing up the deeper water, and the surface near me was calm. As the water reached my shoulders, I let myself melt into the soft sea.

archelon

Visibility was not bad, a lot like it is in the waters of Southeast Alaska in summer when the plankton flashes like dust in the sun filtering down. I hovered at forty feet, where I could see at least that far all around me. The creatures of the Niobrara snapped into view and just as suddenly disappeared into the surrounding darkness, like the jump-up monsters on a carnival fun house ride. *Xiphactinus* in the flesh is a very scary herring, like a tarpon with the head of a bulldog. Even though it couldn't see me, I almost gagged on my mouthpiece when one gave me a head-on look at its teeth and gullet, the last view of life for many a *Gillicus*. A *Bananogamius* ghosted past, a three-foot-long fish with a fluttering corona of spiny fins, then a saurodon, one of the protosphyraenids—lean and mean billfish, some with their bills on their lower jaws.

I swam a few feet, until I was surrounded by an enormous school of squid, looking just about like they do now, deceptively placid invertebrates. They scattered in a heartbeat as an enormous bird, *Hes-*

perornis, drove through them, trailing bubbles in its rush from the surface, seizing a squid with its toothed beak on the way down. A lurking gar ambushed some of the fleeing squid, and a *Xiphactinus* joined the frenzy. The squid dispersed and, as though to replace them, a flotilla of ammonites filed by, looking like jet-powered snare drums, some of them three feet in diameter. Their eyes were startling—alert and searching—carrying some hint of the processes we call intelligence. Then, as though in an eclipse, the shadow of an archelon passing overhead stole the light. The protection of the giant turtle's hard shell as well as its size seemed to permit a more tranquil demeanor than those of the other denizens of the Niobrara. Of all the creatures, even the giant reptiles, archelon seemed to show the least fear in this world of slashing teeth, fins, and tails.

The turtle moved away, the light returned, and an enormous creature formed in the periphery of my vision at the very edge of my face mask. I turned and thought, "Nessie," and watched a long-neck plesiosaur, *Elasmosaurus*, undulate as it rowed its enormous body through the water. I looked up and saw the bellies of other elasmosaurs and shook in terror as one of them plunged its head at the end of its twenty-foot neck to snap up an unlucky *Cimolichthys*—a slender, snakelike fish—no more than a couple of arm's lengths from me. I could see the red of its throat and then the blood of the struggling fish. Sharks, naturally, came to mind, and then they were all around me, one of them bigger than any great white in a fright movie, all as threatening as their modern cousins.

Enough, I thought, and I headed for the surface, where I broke water like a wide-eyed refugee smack in the midst of a raft of nyctosaurs—the flying night lizards—bobbing on the gentle swell.

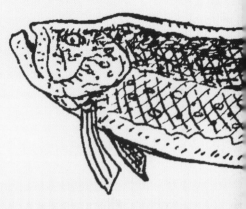

GILLICUS

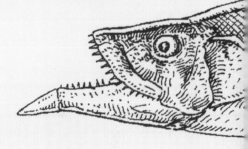

SAURODON

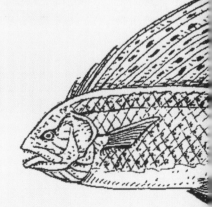

BANANOGAMIUS

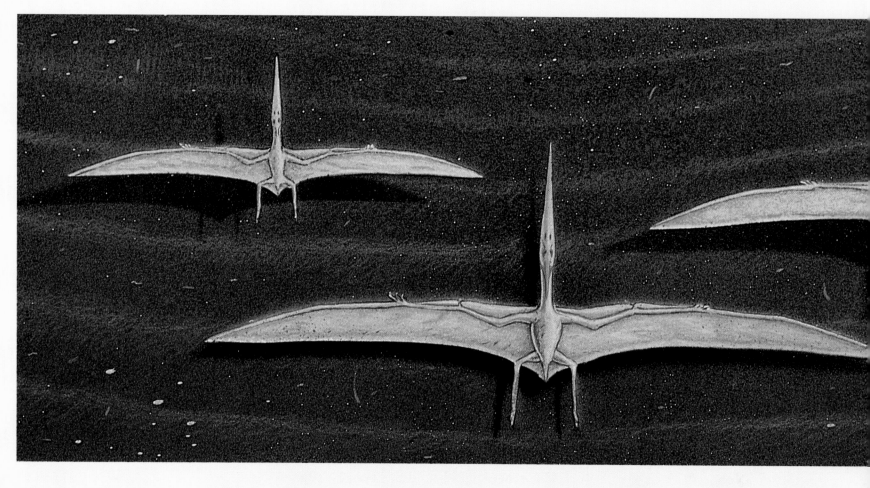

Pteranodon Formation

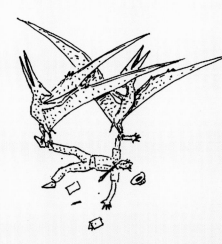

A squadron of others soared overhead against a Magritte sky, draped like canvas kites on the air itself. They cackled and howled at each other and performed slow-motion aerobatics like hang gliders riding thermals. In the middle distance, I could see the roiling mosasaurs, but when I raised my mask for a better look, Kansas was back.

"Ray," I whispered. I was gripping the steering wheel so hard my fingers hurt. "Ray."

"Yo," Ray said, dropping quickly from sleep into reality. "We there?" One of the best things about Ray is that hallucinations are well within his comfort zone, so I had no trouble telling him about

the southbound mosasaurs and my excursion into the Niobrara. "What color were the mosasaurs?" was the first thing he asked, and I told him blackish-green and brown, with stripes like the parr marks on chum salmon, and yellowish-white undersides. And then we were climbing the long grade to the Church of the Cretaceous, and Ray gave up reading about mosasaurs because the light was gone.

At the Bonners', before we went into the house, we stood around and watched the moon rise and listened to the whack-whack of the windmill and meandered through the delicacies of strangers about to spend a couple of days together. Chuck,

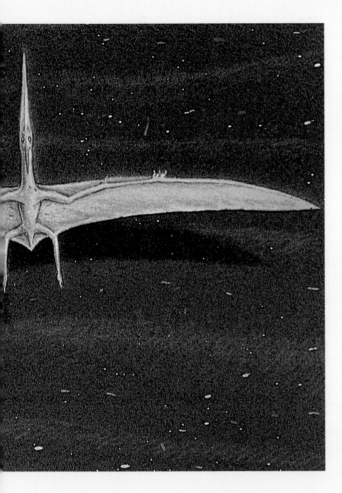

A squadron of pteranodons soared overhead against a Magritte sky, draped like canvas kites on the air itself.

was about the age of his own son, Logan. "I've got a video of Dad out in the beds," Chuck told us. "I just can't watch it yet. I'll send it to you." His father had been mightily amused by the hubris of humankind and, with the gleam of wisdom in his eyes after eight decades in the Mesozoic, he liked to describe himself as "just another vertebrate." So we drank a toast to Marion Bonner, arguably the last of the old-time fossil hunters, and turned to the business of settling our minds before sleep.

The full moon at midnight bathed the prairie with silver light, and I took a walk around the place before risking the dew in my bedroll under the clear Kansas sky. Chuck, Barb, and Logan live in a house built from blocks of Cretaceous limestone, formerly the parsonage of the Pilgrims' Holiness Church of Pleasant Ridge, founded in 1917. The church itself is now their art and fossil gallery, and the town of Keystone, which was nearby, no longer exists. The Bonners are just about self-sufficient, drawing power from a windmill, water from a well, vegetables from a garden, meat by bartering with a neighbor, and what little cash they need from their artwork and the fossils that surround them. Logan found a couple of knuckles of a *Pteranodon* when he was three and a fine mosasaur skull when he was five. Barb seems to have good luck finding fish, like the near-perfect *Xiphactinus* skull they now display in the gallery. Chuck has found and prepared some of the best recent specimens of late Cretaceous marine reptiles, flying reptiles, fish, crinoids, and giant clams in the world. Like his father, though, he maintains a modest profile.

The third building on the Bonner place is a

Barb, Ray, and I faced each other like we were playing ring-around-the-rosy, shifting from leg to leg and crossing our arms against the evening chill. Their eleven-year-old son, Logan, circled us like a herding dog. The breeze freshened. When Chuck said he'd seen the wind so strong on this hill where he lives that it blew the cheese off the cheeseburgers on his barbecue, our laughter ended the preliminaries. Fossil lust is solid common ground, so we were easy with each other. Then Chuck led us into the kitchen of his stone house and showed us pictures of himself, his brother Orville, and his father digging out another good plesiosaur in 1955 when Chuck

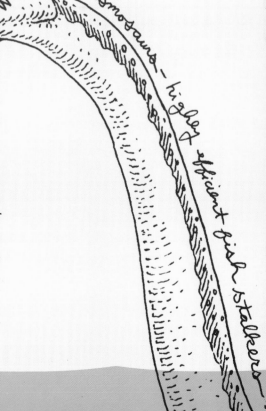

elasmosaurs – highly efficient fish stalkers

Death at sea for a pterosaur included sinking to the ooze of the sea bottom through a circus of hungry mosasaurs, plesiosaurs, and turtles.

shed, the inevitable "put-it-out-back-for-now" shack for whatever needs storage for later study or use. I wandered over to where the trimmed grass of domesticity gives way to the rough scrub of the prairie and found Chuck's mosasaur corral. Penned by a chicken-wire fence against the back of the shed was another herd of mosasaurs, waiting for preparation. They were laid out on plywood sheets on sawhorses, their bones askew in the usual jumble of raw fossils, still in the chalk matrix that carried them through 65 million years to the Church of the Cretaceous here at Keystone. Finally, I was in my sleeping bag, ready for the dreamworld again, knowing that my mind would pull its night tales from Chuck's corral, that I would see the mosasaurs and the Niobrara again as I had driving that afternoon. I settled to rest, with the *Pteranodons*, on the languid surface of the Cretaceous sea.

ADRIFT ON THE NIOBRARA

THE NEXT MORNING in the truck on the way to the beds with Chuck, Barb, and Logan, I told Ray about the pterosaurs landing on the water, and he asked me if I could see what they did with their wings. Paleontologists shy away from speculating that avian reptiles—pterosaurs—landed on the water because they can't figure out the wing loading or musculature necessary to get them into the air again. Unlike artists, few want to risk their scientific reputations on a bad guess. But just imagining how these creatures folded their enormous wings was a tall order,

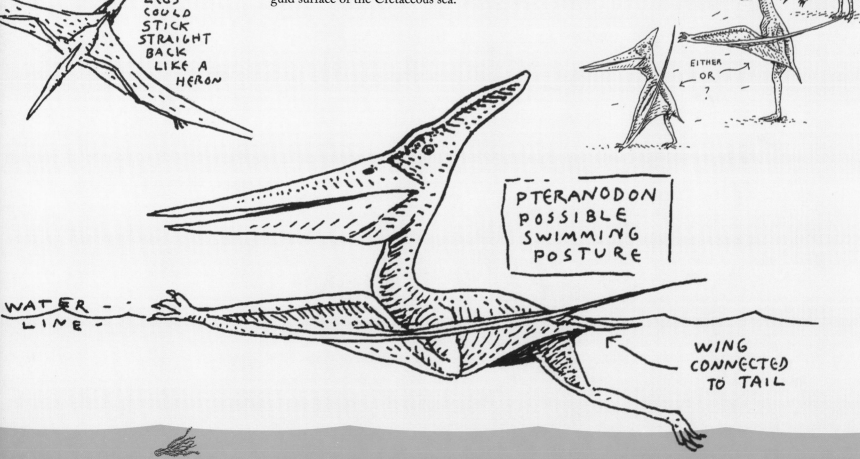

LEGS COULD STICK STRAIGHT BACK LIKE A HERON

OR

EITHER OR ?

PTERANODON POSSIBLE SWIMMING POSTURE

WATER LINE

WING CONNECTED TO TAIL

since some of them had spans of over thirty feet. A lot of people think that pterosaurs swooped over the sea and scooped up their food from the surface, never landing at all until they returned to their highland perches.

During the Cretaceous, though, the land nearest the beds now at Keystone was several hundred kilometers away. Kansas and the Niobrara Sea are famous for pterosaurs, and the two species of the *Pteranodon* genus are local to the Smoky Hill chalk. In one of the opening volleys in his fossil war with Cope, one of O. C. Marsh's minions found the very first one in 1870 just a few miles from Keystone. The other pterosaur genus from the Niobrara is *Nyctosaurus*, a smaller model so rare as a fossil that less than thirty have been found, compared with hundreds of *Pteranodons*. Death at sea for a pterosaur included sinking to the ooze of the sea bottom through a circus of hungry mosasaurs, plesiosaurs, and turtles. Only fragments of a smaller, bite-sized morsel like a *Nyctosaurus*, with its six- or seven-foot wingspan, would make it all the way down. The true swimming birds of the late Mesozoic, like *Hesperornis*, suffered the same fate, and, though occasionally found in the chalk, they are also rare.

You almost never find the fossil of a whole animal, but you do have to find enough to be able to identify it. In museum exhibits open to the public, the skeletons are often far less fossil than reconstruction, though specimens available to scientists in the back rooms are almost never doctored up. The Bonner family name for adding missing bones is "weenie-ribbing," a mildly pejorative term as in "That *Xiphactinus* in Lawrence is mostly weenie-ribbed." You can weenie-rib a fossil from very small scraps, which is why sharp eyes in the beds can produce great discoveries right on the surface and why

you hunt fossils in the morning and evening when the flat light brings contrast to the glare of the chalk.

We had risen at dawn, too, because the heat was sure to be unbearable by noon. Chuck had permission from nearby landowners who used the range for grazing livestock to hunt fossils, and we didn't have far to go. Bustling around in dim light, Barb made us a lunch of bone-hunting food—peanut butter and jelly sandwiches, potato chips, coffee, and water. Everybody talked about the glorious discoveries of the past and expressed high hopes for the day. "Remember that *Pteranodon* you found, Logan," Barb said, rattling around the near-dark kitchen. "You dug it out and it smelled like oil." And Logan told his story about finding a mosasaur skull when he was five years old: "I yelled, 'Dad, I found something,' and brought a piece of the jaw over, and sure enough it was a mosasaur. It looked just like a gar jaw except bigger with fewer teeth. That's how I knew."

Chuck talked about his fever for giant clams in recent years, a foray into invertebrates that has produced some fine specimens—one in particular, which went to the Natural History Museum of Los Angeles County. The museum hired J. D. Stewart, a Kansas-bred vertebrate paleontologist from the line of Sternberg, Walker, and Marsh. Somebody in Los Angeles finally realized they had catacombs full of specimens from the chalk, many of them found by the Bonners, that needed identification and preparation. Chuck found one of them in the mid-eighties while he was playing golf out in the chalk. "I was hitting nine irons, lost all my balls, and when I went looking for them found a real nice *Trinacromerum*," he said. He and his father dug it out and shipped it to Los Angeles, where it reposed in the basement until J. D. took a good look at it and realized that a juvenile or embryonic plesiosaur was inside the

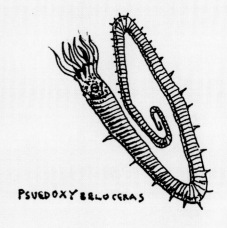

PSUEDOXYBELOCERAS

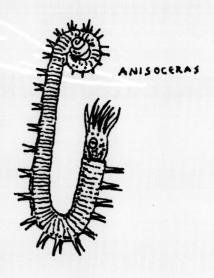

ANISOCERAS

We took turns lying down in the mosasaur quarry, Chuck first, then Ray, then me. I didn't want to get up. I looked up into the evening sky over Kansas and the first breaths of starlight, felt the imprint of the mosasaur in the rock against my back, and heard the whistling of the Pteranodons.

Raymond Ptroll:
***Pteranodon* Hunter**

parent, as Chuck had mentioned when he found the fossil. The treasure is still in a crate, but at least paleontologists know where to find it.

Chuck Bonner's family nicknamed him "Coyote" because he skittered around the rest of them to hunt alone, his head bowed to scan the terrain underfoot in the fossil hunter's slouch, alert for the slightest ancient signal of a fossil. Logan hunts that way now, while Chuck, Barb, Ray, and I tend to stay in a pack to talk about what we see—a fish bone here, a chip of something unidentifiable there. With uncanny certainty, Chuck spotted what he thought was a scrap of mosasaur in the slope of a yellowish chalk hummock and dove to his knees, scraping and brushing almost before he hit the ground. The Bonners, like all field veterans, cherish their tools, such as the hard-to-find Marsh Picks, lightweight, single-hand chipping hammers with flat, pointed prongs for excavating loose marl and rock. A good four-inch paintbrush is standard for clearing away debris as you work. And every collector has handmade scrapers, usually wooden-handled chisel and awl hybrids, for the fine work on a fossil that could prove priceless. "You never know what you've got until you dig on it," as the saying goes, and that means until you know otherwise, every partially exposed sliver of mosasaur or *Pteranodon* must be explored as though it were a perfect, fully articulated specimen. "Most of the time they quit on you," Chuck said, brushing away at this hint of a mosasaur, eventually pronouncing it a "maybe" and deciding he'd come back later to dig some more. Most of the time, a large fossil requires quarrying, a chore that is every bit as much fun as hard-rock mining.

"I like the hunt, but I don't much care for quarrying," Chuck said, becoming Coyote again. The rest of us fanned out, excited by the maybe-mosasaur just as fishermen are by a good, hard strike—even when the fish gets away. As the morning unwound, walking the chalk became more and more like the banquet of fishing. Anticipation was the main course and telling of fossils-past the appetizer. Chuck said experiments are under way to use density detectors for finding fossils, and his voice seemed to fall as he contemplated replacing the thrill of the hunt with the calm certainty of technology. "Big dinosaurs are worth millions," he ventured. "People want to get to them any way they can."

Suddenly, Ray yelled the paleoequivalent of "Fish on," something like "Hey, Chuck, what's this?" He stood and pointed at the slope of the steep knob we were circling. Everybody crowded around for a look, leaning into the hill. There we saw a pencil-thin piece of something darker than the chalk, darker, too, than the common fish fossils we had learned to recognize in the two hours that morning. Again, Chuck dove to his knees, brushing and scraping, this time talking to himself with a different tone than he used on the maybe-mosasaur. It was more like a delighted clucking: "Yep, yep, that goes on in there, yep, yep. I guess, I can't be sure, I guess what you've got here is the knuckle bone of a pterosaur, probably a *Nyctosaurus*." Ray grinned and got right down on the chalk next to Chuck, who was quickly exposing a good six inches of bone that was clearly something, even to a novice. Finally, Chuck pronounced this a strong maybe for a good *Nyctosaurus*. If it panned out after quarrying, it would be only the twenty-eighth specimen ever found. And it did, six months later.

The ninety-degree heat was curdling my brain by ten that morning when we quit, but I'd swear Raymond Ptroll, *Pteranodon* Hunter, was floating above the chalk on our way back to the truck, with

graphapithecus matsenni

flocks of the reptiles soaring overhead, and the mosasaurs and plesiosaurs rising into the air, once again, this time to applaud the rookie from Alaska who found a great fossil. Ray found the *Nyctosaurus* again that evening, too. Around the barbecue grill as Chuck cooked steaks in the wind, we talked *Pteranodon* like fishermen on the beach talking about a big steelhead and a good day on the water. After dinner, the light flattened and hung over the prairie like an exhale. Chuck took us to another of his favorite beds to show us a quarry he'd dug to get out a mosasaur. This time, we bounced off the road for a few miles in the pickup, finally dismounting on a plateau that overlooked a broad scar in the terrain.

While the canyons and hoodoos of the morning hunt gave the impression of choppy water, this place looked more like rolling combers, much less abrupt, with long crests and troughs. On denuded ridges, we walked across exposed slabs of rough fossil shells, remnants of oyster beds and the Cretaceous ooze on the bottom of the Niobrara that still smelled like the sea, Chuck told us, after a rain. His quarry was a ten-by-ten-by-ten-foot cave hacked into the slope of one of the "waves," from which he'd excavated a nice *Clidastes*, now in his corral awaiting preparation. All day, we'd talked off and on about S. P. Dinsmoor and the Garden of Eden, and Chuck's dad, and about people trying to bring meaning to their lives and deaths. We decided that one of the best things you could do was to get yourself situated right before you die so you make a good-looking fossil. We talked about creation and about how Marion Bonner knew he was just another vertebrate, instead of some crowning glory of divine intention. Chuck said, that week after his father's death, "Some people ask, 'Why don't natural history museums have creation sections?' and I ask,

'Why don't churches have fossil sections?'" We laughed so hard there on the ridge we almost fell down, and we took turns lying down in the mosasaur quarry, Chuck first, then Ray, then me. I didn't want to get up. I looked up into the evening sky over Kansas and the first breaths of starlight, felt the imprint of the mosasaur in the rock against my back, and heard the whistling of the *Pteranodons* as they settled for another night.

OH NO, NOT AGAIN

THE MESOZOIC WOUND DOWN and probably ended with a stroke of the cosmic hammer 65 million years ago. Almost certainly, though, before a comet or asteroid administered the coup de grâce, the species diversity that flourished in the Triassic and Jurassic had decayed by the time the dark moment arrived. Highly evolved vertebrates that eat all the time are particularly hard on Planet Ocean. By the end of the Cretaceous, the tribes of dinosaurs, pterosaurs, and great marine reptiles and their prey—including the long-running ammonites, some of the fish, and any large birds or upstart mammals that couldn't hide—were very vulnerable. Evolutionary voltage eventually trickles out of a gene pool as it ages over tens of millions of years. And then there is the simple matter that extinctions are as natural as gurgling babies.

The small, meek, and well-hidden inherited the debris on the other side of the Cretaceous extinction and bloomed again within just a few million years. Mammals, fish, reptiles, and birds—whether they are distant cousins to dinosaurs or not—carried the flag for the vertebrate tribe, and hordes of arthropods, our insect friends, took over as the main event. No new tribes have come ashore, though members of all have returned to the sea, once again. And now, 65 million years into Planet Ocean's latest refinement of the passenger list, *Homo sapiens sapiens* is looking it over and saying, "Hold on there. The next mass dying seems to be in the works." Extinction's drum is sounding somber notes, and we're at the top of the list if you look at the criteria of size, specialization, and reproductive vulnerability.

We're not quite sure what effect the new wrinkles of intelligence and self-awareness have on the prospects for our own survival or that of the 50 million other species trying to make a living on a crowded globe. Clearly, the proliferation of such consumptive beings is wearing on the abilities of the other tribes to survive a mass dying. The balance seems to have been disturbed, though, because intelligence just happens to have been visited on a carnivorous primate instead of some other animals. But who knows? Perhaps we, like the large vertebrates that set up the Permian, Cretaceous, and other extinctions, are simply the straws that stir the drink of life every once in a while.

It would have been hard to devise a critter more suited for high-volume destruction of energy than *Homo sapiens sapiens*. In his fine essay, "Is Humanity Suicidal?" (*New York Times* Magazine, May 1993), E. O. Wilson struggles, as Ray and I did between the trout stream and the fossil beds, to understand how the ecosystem of Planet Ocean can remain in balance with human intelligence in the equation. "The human species is, in a word, an environmental abnormality. It is possible that intelligence in the wrong kind of species was foreordained to be a fatal combination for the biosphere. Perhaps the law of evolution is that intelligence usually extinguishes itself."

All this can really worry beings who have been raised to see themselves as universal treasure and who have the intelligence to presume control of the biosphere. The big question for most human beings is not "How can I revise my daily vision of harmony to incorporate death as an element of survival of life on earth?" but "How can I control this carnival ride of evolution so death doesn't happen, at least to me and my kind?" One camp reacts to the new uncertainties of life on Planet Ocean by insisting that our intelligence itself will allow us to survive through what Wilson calls "scientific and entrepreneurial genius," curing the ills of a declining biosphere,

SLOW TROLL ON THE NIOBRARA

as though it were a backyard garden. The other camp, loosely classed these days as environmentalists, is slowly, finally coming around to a vision of balance in which humans are not the main event. As we head for another mass dying, we finally know any shot at the survival of life at all depends entirely on preserving diversity and strength, not only in ourselves but in as wide a variety of species as possible. At the same time, we have to put the brakes on the population explosion of the wise, wise beings. If humans lay waste the ability of other species to survive, the next extinction could eclipse even the Permian in its cleansing of life from Planet Ocean.

Paleo Fishing Charters

THE CENOZOIC

U-PICK 'EM IN THE EOCENE

FISH AND MAMMALS SNUCK through the Cretaceous extinction and a good thing, too, because this book begins on a steelhead creek. We wouldn't have gotten to page one if our relatives had vanished with the dinosaurs, great marine reptiles, and all those other plants and animals. But the ash settled and the sky cleared after a few hundred thousand years, and, as usual, the survivors had a ball with the scoured seas and continents of Planet Ocean. For 64 million years now, we've been tripping through the Cenozoic—literally, "recent life"—to the beat of the same tune that hums to us from the fossils: Life is uncertain, but it keeps on keeping on.

The era was mapped by the great geologist Charles Lyell, who, despite his religioscientific notions that the earth and its passenger list are as they always were, convinced Charles Darwin that

there has been plenty of time to account for an evolution of species. Lyell laid out the first part of the Cenozoic according to the patterns of mollusk fossils he found along the modern Atlantic Ocean. His conclusions were wrong and too tedious to run through here, but he left us with Paleocene, Eocene, Oligocene, Miocene, Pliocene, and, as named by other geologists, the Pleistocene and Holocene. (Paleontologists Eat Only Murky Plankton Porridge Hot, remember?) In the recovering Cenozoic, life again fashioned economies of energy production and consumption, continuing to arrive, evolve, and depart as it does at this very moment.

Ray and I were no longer rookie time travelers when we fetched up in the Eocene at Ulrich's Fossil Gallery and U-Pick 'Em Quarry. *Eo* means dawn, and, fittingly, our guide met us in the parking lot just as the morning light broke over Wyoming, tinkering with shadows that sent the bluffs of Fossil

> *"People finally don't have much affection for questions, especially one so leprous as the apparent lack of a fair system of rewards and punishments on earth."*
>
> —JIM HARRISON,
> *Legends of the Fall*

Butte soaring into the brightening sky. Carl and Shirley Ulrich run their gallery, an international brokerage, and fossil charters, and Ray and I paid them $40 each for a trip into the quarry they lease from the state. For the money, we got a ride up a mountain across the valley from Fossil Butte, the tools for the job, and a guarantee that we would find the exquisite fossils of fish that lived during the Eocene in a vast, shifting chain of lakes now named Uinta, Gosiute, Flagstaff, and Fossil.

In that old living-and-dying pattern so fertile for evolution, the lakes filled and dried many times over 2 million years in an area we now call western Wyoming and Colorado, northeastern Utah, and southeastern Idaho. On their shores, in dense fern and flowering forests, birds as big as backhoes were the top critters in the food web for a while, but the great flightless beasts gave way to the most successful carnivores of all time, us. In the Eocene, mammals got a firm grip on Planet Ocean. *Uintatherium* the size of modern bush elephants roamed in herds, and the first hoofed animals brought us a step closer to thoroughbred horse racing and prime rib. A bloom of diversity among rodents fed the large meat eaters, and the first members of our order, Primates, began the unique evolutionary journey toward self-awareness, mass consumption, psychotherapy, religion, war, science, home ownership, and Big Macs. The final grades are still not in on shortened snouts, hands, feet, large brains, heavy equipment, and a puzzling tendency to foul the nest.

We felt a little cheesy about paying for fossil hunting. Ray had already passed the Honorary Paleontologist test at the Royal Tyrrell and found a

fine *Nyctosaurus* in the Niobrara chalk, and I had done a lot of reading and, mercifully, managed to find a plesiosaur's paddle bone. The truth is, though, we were crazed over the long-shot chance that we might find a gar, our old, inspiring signal of survival and, after the sand pit on the Arkansas River with Steve and Scotty, our good luck token. Fossil gar are rare, but the Ulrichs' son, Wally, found the biggest one ever when he was ten years old. The five feet nine inches of gorgeous fossil *Lepisosteus* displayed in the place of honor in the gallery sports the same toothed jaws as the skull we duct-taped to the bumper of our truck in Kansas. Our gar lust flattened out a bit when our guide, a fresh-faced geology student from Wisconsin named Mike, told us we had to hand over to the state any big-time fossils we found.

Carl and Shirley Ulrich grew up in Kemmerer, the hardscrabble mining town ten miles east of where they live now, the abandoned railroad town named Fossil, founded in 1880. They married in 1947 and started wandering around the Green River hardpan with a wheelbarrow because collecting fossils was free entertainment and they were young and broke. Later, they turned pro, and because Carl is a master preparator and some of the richest fossil beds in the world are nearby, their fossils are in museums and private collections all over the world. "We raised our family in the quarries," Shirley says, her voice carrying the overtones of deep satisfaction and a trace of self-promotion.

Up on the mountain, Ray, Mike, and I slipped flat iron bars into cracks in the mudstone and pried slabs of life from the Eocene. The morning light revealed subtle ripples that could have been the

We looked down at the fragments of scales as big as silver dollars, frozen in the hard Miocene rock, the remains of a giant fish with relatives still swimming in the Pacific.

OPPOSITE: Saga of the Sabertooth Salmon

**LAND AND SEA DURING
THE MIOCENE**
5 to 25 million years BP

**The continents looked about
like they do today, but Planet
Ocean was still shuffling the
tectonic deck. Africa collided
with Europe, and up jumped
the Alps. India piled into Asia
to make the Himalayas. The
cast on land included the first
pelicans, pigeons, parrots,
woodpeckers, apes, mice,
rats, falcons, finches, crows,
kangaroos, horses, giraffes,
and upright vertebrates a lot
like us.**

backbones of fish, looking as though they were still swimming just under the surface of the rock. The fish of the Eocene lakes and oceans were remarkably similar to those in 1993, a marvelously diverse collection of prey and predators. The most common are the herringlike *Knightia* and *Diplomystus*, larger perch named *Mioplosus* and *Phareodus*, and *Priscarara*, the bass-like fish that could be held responsible for all those TV fishing shows with sports stars, bassbabble, and bad music. I half expected we would turn up a fossil baseball hat with rod and reel manufacturer's name on the crown.

The freshwater bass are way-cool, as Ray says, but the saltwater sediments of the Cenozoic tease anglers with even more astonishing variations on the fish theme. Later, we got a look at salmon-to-die-for one afternoon in Los Angeles when J. D. Stewart, promising us a surprise, led us down an aisle of specimen cabinets in the basement of his museum. Ray and I were salivating while J. D. talked on about the disastrous economics of southern California and the firing of many of the museum's curators to reduce costs. Somebody had taped a bleak notice to the wall: "Will Curate for Food." At the end of the aisle, J. D. finally stopped and pulled open a drawer. "Here you go," he said. "*Onchorhynchus rastrosus*—a sabertooth salmon from the late Miocene, 10 or 12 million years ago." We were quivering as he handed Ray a skull that could have come from a modern coho salmon. Until we looked closely, it was nothing to frighten even the most timid wading angler. But then we saw the fangs. "This is the small one, a juvenile we think," J. D. said. "But check this out." He pulled open another drawer. "This was found on a cliffside at Jalama Beach, California, between Point Conception and Point Arguello in 1963 by Irving R. Neder,"

J. D. recited from the specimen card. "I found it in the collections, just looking around one day. This salmon was eight feet long in life." We looked down at the fragments of scales as big as silver dollars, frozen in the hard Miocene rock, the remains of a giant fish with relatives still swimming in the Pacific and spawning in watersheds from California to Japan. "How'd you like to hook this one?" J. D. asked.

Meanwhile, back in Fossil Butte, Mike told us we should be alert to a kind of altered reality once we got going with our pry bars. "It's a lot like Christmas morning when you're a kid and you sort of lose consciousness opening presents. That's what this is going to be like, so when I tell you it's time to go, you have to stop right then and don't try to fight me. I've seen people do some pretty strange things to try to stay up here." He was right, and Ray and I were pathetic. Mike showed us how to work the irons into cracks in the sediments, working down in half-inch slabs. Then he leaned back against the Suburban's tailgate and watched our pleasure. Driving the irons into the cracks in the sediment hurts your hands, and lifting them is plain hard labor. In a couple of minutes, though, we found our first definite fossil, and then we were prying and flipping slabs and crying out like madmen when the soft, damp rock broke open to expose the brown remains of the Eocene fish to the light of the Holocene sun.

The mudstone was wet from seepage, and the moisture evoked the aromas of the ancient lakes, musty odors tinged with the biological residues of palm, cinnamon, maple, and oak forests and the crocodiles, turtles, flamingoes, snakes, lizards, fish, horses, rodents, and insects that set the course for life in the Cenozoic. Our two hours flashed by (another kind of lesson in time traveling), and Mike

had to tell us more than once that we had to leave, never abandoning the polite tones of his midwestern upbringing. We found dozens of *Knightia*, and our catch of the day was a nice *Phareodus*, but no gar. Mike said a clump of tangled bones in one slab might be a gar fragment, but he was probably being nice to the customers. As we bounced our way to the valley floor on the steep two-track, Mike commented, ever so gently, on our embarrassing display of fossil lust. "I told you so," he said. We just grinned.

Back in the parking lot at the gallery, Mike shaped our fossils into neat squares and rectangles with a bandsaw, while Ray and I talked to Wally Ulrich, gar finder, poet, and master preparator. One of his specimens, a seven-by-fourteen-foot slab carrying a spectacular palm frond, took him 1,000 hours to prepare and is now part of the permanent collection at the Smithsonian, an object of pride to Wally. Like many of the paleontologists we met, the Ulrichs convey a deep, clear sense of their place on Planet Ocean and an ease with environmental ethics that remove humans from the starring role. Carl will tell you, if you probe a bit, that his fascination with fossils makes him more aware of what's living here on earth. Wally is a well-known environmental activist and also subscribes to the theories of Ben Jepson of Princeton, who believes human beings are something of an evolutionary mistake. "Paleontology has enormous social significance," Wally told us while we loaded our fish to leave. We had two boxes of fossils, mostly *Knightia*, wrapped in newspaper to absorb the moisture and protect their fragile songs from the past. "An involvement with earth science immerses us in a tactile relationship with time and the earth," he said, slipping into the fervor of a messianic preacher,

SMILODONICHTHYS (ONCORHYNCHUS) RASTROSUS, HOMO SAPIENS SAPIENS, AND ONCORHYNCHUS TSHAWYTSCHA

waving his arms around, with Fossil Butte rising like a prop behind him into the hot, high-country afternoon. "I always wonder why the creationists think the ways of God are so limited. There is almost certainly no force directing life, but something's going on here that we can't understand. The earth will continue to go through the cycles we see in the fossils from being a very rich planet to being a very poor planet. Humans are threatened now, that's for sure, and we have to step lightly. But I don't have to know what happens next to be responsible for my presence on earth. To me, uncertainty is the greatest argument for environmentalism."

©RAY TROLL 1992

CODA

STILL UP THE CREEK

WHEN I STARTED WRITING *Planet Ocean*, I was sure the story would circle neatly back to Fish Creek. I thought I could make sense of evolution by telling you the steelhead was foul-hooked and died so we ate it that night over a fire in front of our cabin, then lay in the dark a little drunk in our sleeping bags, rustling around on the hard wooden bunks and telling stories about how good fishing used to be. I thought something about our biological relationship with that fish was the theme of *Planet Ocean*, and, in a sense, I suppose it is. We are more fish than not. But I intended to artificially stretch our trip through time between the moments when the fish was hooked and the fish died, literary sleight of hand that would have represented life as far too orderly and conclusive.

We did watch that steelhead die on Fish Creek two winters ago, but when Ray and I really ended our story, we were crawling around like babies on a Devonian coral reef, looking for trilobites with a few good friends on a bright June morning. The reef had formed, thrived, and died more than 350 million years ago and, during the tectonic rhumba that created Southeast Alaska from oceanic and continental crust, became part of Kasaan Island. The little island is a piece of the geologic puzzle wildly out of place, a hunk of limestone that should be a hundred miles west of where it is. We shed our raingear, sweaters, and heavy shirts to expose ourselves to the warmth and light of the beautiful day and reveled in knowing that the coarse old coral roughing up our hands and knees had been alive at about the same time that that first hardy fish dragged itself ashore. The 350-million-year-old reef repeats the undeniable truth that powerful streaks of life thrive for millions of years, and then some are gone and some are not.

We stood transfixed, looking up at the bottom of the Devonian Ocean, and knew steelhead streams would never again exist only in the present.

OPPOSITE: Mandala

Delight seeped from us in giddy chatter and bad jokes. Getting up at dawn, packing lunches, and drinking strong coffee on a boat ride reminded us of fishing trips, so we got a lot of laughs just switching "trilobite" for "fish" in the jargon of angling. "It's real hard to tell a trilobite strike from just dragging bottom. Don't horse it if you get one on." "How deep are you?" "I just throw them on the grill with garlic, lemon, and brown sugar." And on and on, while we cackled and hunkered over the ancient seafloor. Each of us put a dollar into a hat, the prize for the first person to find a trilobite.

One of the boat's crew won the pool with a nice, though weathered, trilobite. Each of the rest of us found at least one in the four hours we were ashore that day. Ray and Jim Baichtel, the only real geologist on the trip, lucked into the concave and convex halves of the same trilobite a few feet apart, a wonderful moment eclipsed in my memory only by what we saw just before leaving for town. Jim, who had been to Kasaan, led us up into the forest to a rock outcropping hidden from the water, pocked on its seaward side by caves and undercut along a thirty-foot face to form an overhang shelter. The white limestone was darkened at intervals by rising streaks of soot, campfire sites that signaled the presence of *Homo sapiens sapiens*, the fire-builders. Near each charred site, piles of rotting firewood testified to thought and preparation. In places, long logs remained tipped against the overhang to form what must have been welcome protection against the wind and rain hundreds of years ago when bands of the early humans crowded here for shelter.

They had found a good spot, I thought, with plenty to eat from the sea and nearby shoals of berries and other edible plants in the forest. They wouldn't have known that the wavy lines in the dark rock where they gathered mussels were the remains of impossibly old coral, or that the cliff against which they cooked and crouched at night was once a cloud of organic debris settling to the seafloor. When they chanced to follow the sparks of their fires up to the overhanging limestone, they would not have known, as we did, that the odd circles were the bottoms of stromatolites, the first colonies of complex life. We stood transfixed, looking up at the bottom of the Devonian Ocean, and knew steelhead streams would never again exist only in the present. I struggled for a while to find reassuring words for an ending full of answers, but finally I realized the answers are not in a neat ending on a steelhead stream, that joy and harmony are in the search.

FIN

© RAY TROLL 1994

RELATED READING

GEOLOGY AND PLATE TECTONICS

Alt, David D., and Donald W. Hyndman. *Roadside Geology of Washington*. Missoula: Mountain Press, 1984.

Burchfield, Joe D. *Lord Kelvin and the Age of the Earth*. New York: Science History Publications, 1975.

Connor, Cathy, and Daniel O'Haire. *Roadside Geology of Alaska*. Missoula: Mountain Press, 1988.

Cox, Allan, et al. *Plate Tectonics and Geomagnetic Reversals*. San Francisco: W. H. Freeman, 1973.

Greene, Mott T. *Geology in the Nineteenth Century*. Ithaca: Cornell University Press, 1982.

Hallam, A. *A Revolution in the Earth Sciences: From Continental Drift to Plate Tectonics*. Oxford: Clarendon Press, 1973.

Huxley, Thomas Henry. *On a Piece of Chalk*. New York: Charles Scribner's Sons, 1967.

Kiefer, Irene. *Global Jigsaw Puzzle: The Story of Continental Drift*. New York: Atheneum, 1978.

Lambert, David. *The Field Guide to Geology*. New York: Facts on File, 1988.

Langshaw, Rick. *Geology of the Canadian Rockies*. Banff, Alberta: Summerthought Publications, 1989

McPhee, John. *Basin and Range*. New York: Farrar, Straus & Giroux, 1980.

——. *In Suspect Terrain*. New York: Farrar, Straus & Giroux, 1982.

Steno, Nicolaus [Niels Stensen]. *The Earliest Geological Treatise (1667)*. New York: St. Martin's Press, 1958.

Stokes, William Lee, et al. *Introduction to Geology, Physical and Historical*. Englewood Cliffs, N. J.: Prentice-Hall, 1978.

Thompson, Susan J. *A Chronology of Geological Thinking from Antiquity to 1899*. Metuchen, N. J.: Scarecrow Press, 1988.

ICHTHYOLOGY

Moy-Thomas, J. A., and R. S. Miles. *Paleozoic Fishes*. Philadelphia: W. B. Saunders Company, 1971.

Spoczynska, Joy O. I. *An Age of Fishes*. New York: Charles Scribner's Sons, 1976.

Zangerl, Rainer. *Handbook of Paleoichthyology*, Vol. 3A. Stuttgart and New York: Gustav Fischer Verlag, 1981.

NATURAL HISTORY

Berra, Tim M. *Evolution and the Myth of Creationism*. Stanford: Stanford University Press, 1990.

Dawkins, Richard. *The Blind Watchmaker: Why the Evidence of Evolution Reveals a Universe without Design*. New York: W. W. Norton, 1987.

Frank, Louis A., with Patrick Huyghe. *The Big Splash*. New York: Carol Publishing, 1990.

Gleick, James. *Chaos: Making a New Science*. New York: Viking Penguin, 1987.

Hoyle, Fred, and N. C. Wickramasinghe. *Evolution from Space*. London: J. M. Dent, 1981.

——. *Cosmic Life-Force*. New York: Paragon House, 1990.

Overbye, Dennis. *Lonely Hearts of the Cosmos*. New York: HarperCollins, 1991.

Peters, David. *From the Beginning: The Story of Human Evolution*. New York: Morrow Junior Books, 1991.

PALEONTOLOGY

Benton, Michael J. *The Reign of the Reptiles*. New York: Crescent Books, 1990.

British Museum Trustees. *Dinosaurs and Their Living Relatives*. London: British Museum (Natural History), 1979.

Colbert, Edwin H. *The Great Dinosaur Hunters and Their Discoveries*. New York: Dover Publications, 1968.

Crichton, Michael. *Jurassic Park*. New York: Alfred A. Knopf, 1990.

Dixon, Dougal, Barry Cox, R. J. G. Savage, and Brian Gardiner. *The Macmillan Illustrated Encyclopedia of Dinosaurs and Prehistoric Animals*. New York: Macmillan Publishing Company, 1988.

Eldredge, Niles. *Life Pulse: Episodes from the Story of the Fossil Record*. New York: Facts on File, 1987.

Gould, Stephen Jay. *The Flamingo's Smile: Reflections in Natural History*. New York: W. W. Norton, 1985.

——. *Wonderful Life: The Burgess Shale and the Nature of History*. New York: W. W. Norton, 1990.

——. *Bully for Brontosaurus*. New York: W. W. Norton, 1990.

Gould, Stephen Jay, Peter Andrews, John Barber, Michael Benton, Marianne Collins, Christine Janis, Ely Kish, Akio Morishima, John J. Sepkoski, Jr., Christopher Stringer, and Jean-Paul Tibbles, with Steve Cox. *The Book of Life*. New York: W. W. Norton, 1993.

Howard, Robert West. *The Dawnseekers: The First History of American Paleontology*. New York: Harcourt, Brace, Jovanovich, 1975.

Lambert, David. *The Field Guide to Prehistoric Life*. New York: Facts on File, 1985.

Lauber, Patricia, and Douglas Henderson. *Living with Dinosaurs*. New York: Bradbury Press, 1991.

Martin, Larry D. *Sea Monsters of the Midwest*. Lincoln, Nebraska: Bright Child Books, 1990.

McGowan, Christopher. *Dinosaurs, Spitfires, and Sea Dragons*. Cambridge, Massachusetts: Harvard University Press, 1991.

Parker, Steve. *The Practical Paleontologist*. New York: Simon & Schuster, 1990.

Raup, David M. *Extinction: Bad Genes or Bad Luck?* New York: W. W. Norton, 1991.

Raup, David M., and Steven M. Stanley. *Principles of Paleontology*. 2d ed. San Francisco: W. H. Freeman, 1971.

Russell, Dale A. *An Odyssey in Time*. Minocqua, Wis.: NorthWord Press, 1989.

Shor, Elizabeth Noble. *Fossils and Flies: The Life of a Compleat Scientist, Samuel Wendell Williston (1851–1918)*. Norman: University of Oklahoma Press, 1971.

Spoczynska, Joy O. I. *Fossils, A Study in Evolution*. Totowa, N. J.: Rowman & Littlefield, 1971.

Stanley, Steven M. *Extinction*. New York: Scientific American Library. W. H. Freeman, 1987.

——. *Earth and Life Through Time*. New York: W. H. Freeman, 1989.

Steel, Rodney, and Anthony Harvey. *The Encyclopedia of Prehistoric Life*. New York: McGraw-Hill Book Company, 1979.

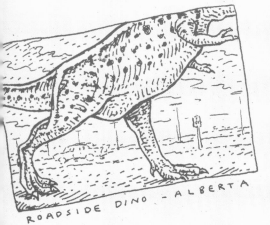

ROADSIDE DINO - ALBERTA

DUST IN THE WIND

FISH OUT OF WATER

Sternberg, Charles H. *The Life of a Fossil Hunter*. Reprint. Bloomington: Indiana University Press, 1990. (Henry Holt & Co., 1909)

Stewart, Ron. *Dinosaurs of the West*. Missoula: Mountain Press, 1988.

Ward, Peter Douglas. *On Methuselah's Trail: Living Fossils and the Great Extinctions*. New York: W. H. Freeman, 1992.

Wellnhofer, Peter. *The Illustrated Encyclopedia of Pterosaurs*. New York: Crescent Books, 1991.

Whitfield, Philip. *From So Simple a Beginning: The Book of Evolution*. New York: Macmillan Publishing Company, 1993.

Wilford, John Noble. *The Riddle of the Dinosaur*. New York: Vintage Books, 1987.

PHYSICS

Davies, Paul, ed. *The New Physics*. Cambridge: Cambridge University Press, 1989.

· N I G H T   L I Z A R D ·

ACKNOWLEDGMENTS

Many people provided us with support, encouragement, and hospitality during the joyful business of writing *Planet Ocean*. They are responsible for most of the pleasure and assistance, and none of the errors. Our thanks to:

Corinna and Patrick Troll; Marlene Blessing; Kate Thompson; Sam Mitnick; Laara Estelle Matsen.

Clem and Diana Tillion; Keith Jefferts and the Fisheries Management Foundation; Howard Hirshman; Colleen Simpson and Tom Webb.

Jim Baichtel; the late Marion Bonner; Chuck, Barbara, and Logan Bonner; Orville Bonner; Andy Neuman; Betsy Nicholls; J.D. Stewart; Peter Douglas Ward; Rainer Zangerl; Grace Hassibar; Chris Bennett; and the late Harold Caldwell.

Lee Alverson; Thomas von Bahr; Monte Dolack and Marybeth Percival; Jan Eddy and Carol Boorady; Kurt Esveldt, Kay Wilson, Rose, and Michael; Barbara Marrett; Stan Moberly; Chip Porter; Ed Reading; Thorne and Marget Smith; Rich Stage; Phil Zeidner and the crew of the *R/V Pacific*; Gary McWilliams; Steve Fairchild; Joan and John Heter; Carl and Wally Ulrich; Bob Widness; Michael Bucove; Doug Charles; and the late Richard Brautigan.

Colleagues at National Fisherman Magazine and Journal Publications.

And, of course, Phil Wood, George Young, David Hinds, and the rest of the fine vertebrates at Ten Speed Press.

Artwork was photographed by Normand Dupre.

INDEX

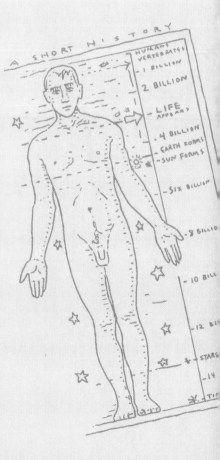

ROAMING
RHIPIDISTIANS
LOOKING FOR A
HOME.

OUR FINEST CATCH THAT D...
PHAREODUS ENCAUSTUS
it looked like a cross between a piranha and a bass with a very bad attitude—